D0857603

SUPERMATH

SUP3R MATH

The Power of Numbers for GOOD and EVIL

ANNA WELTMAN

JOHNS HOPKINS UNIVERSITY PRESS
Baltimore

Printed in the United States of America on acid-free paper
9 8 7 6 5 4 3 2 1

Johns Hopkins University Press
2715 North Charles Street
Baltimore, Maryland 21218-4363
www.press.jhu.edu

Library of Congress Cataloging-in-Publication Data

Names: Weltman, Anna, author.
Title: Supermath : the power of numbers for good and evil / Anna Weltman.
Description: Baltimore : Johns Hopkins University Press, [2020] | Includes bibliographical references and index.
Identifiers: LCCN 2019040486 | ISBN 9781421438191 (hardcover) | ISBN 9781421438207 (ebook)
Subjects: LCSH: Mathematics—Social aspects. | Mathematics—History. | Mathematics—Aesthetics.
Classification: LCC QA10.7 .W42 2020 | DDC 510—dc23
LC record available at https://lccn.loc.gov/2019040486

A catalog record for this book is available from the British Library.

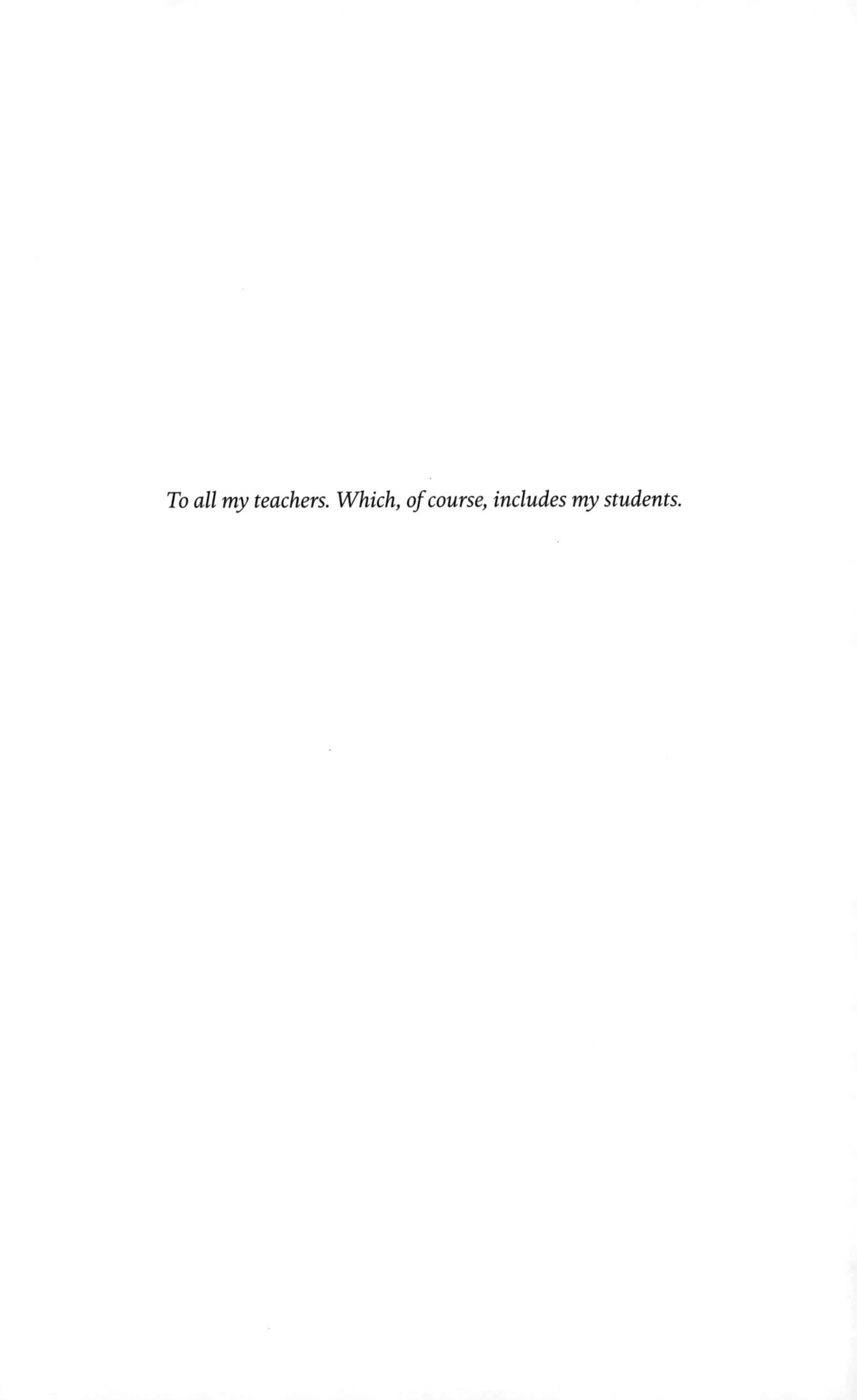

To all my teachers. Which, of course, includes my students.

CONTENTS

PREFACE

A Day at Supermath Headquarters

"You've reached the headquarters of the Amazing Supermath, the one and only mathematical superhero. How can I help you?"

"I found an ancient tablet covered in strange symbols! Can you help me figure out what it says?"

"My district keeps sending Republicans to Congress, but I swear all of my neighbors are Democrats. I think the district might be gerrymandered. Can you help me prove it?"

(In a whisper.) "Hey . . . I'm at the police station. They have me and my friend in for robbing a bank. They don't have enough evidence to charge us, though. They say if I snitch, I can get a reduced sentence. But if my friend also snitches, we're both in for longer. And if I don't snitch and he does, I'm toast. What should I do?"

"When kids from my high school go to a four-year college, they often drop out. They say it's because they keep failing the college's remedial algebra course. But they all took algebra in high school. What's going on?"

"My husband bought this painting at an auction. I think it's ugly, but he thinks it's beautiful. Who's right?"

"I can't do problem number 35 on page 291. Help!"

Math is, at its core, problem-solving. Problems big and small motivate mathematicians to do their work. The importance of problem-solving is what is supposed to make studying math so powerful. As the story goes, if you do well in math in school, you'll excel at any problem-solving job.

But what does "math is problem-solving" really mean?

If you're like a lot of people whose experience with math began and ended in school, most of the math problems you did were probably like problem number 35 on page 291. You might have had a real problem if you couldn't figure out how to solve problem number 35 on page 291 and needed to get it right to not fail your homework assignment. But the math problem itself, whether it was dividing two fractions or figuring out when two hypothetical trains will pass each other along a stretch of track, probably didn't feel like a real problem.

Maybe the problem's context came with a sense of urgency. If you don't simplify the algebraic expression properly, the trains will crash! But even a compelling context rarely captures a real need for math—either because the math demanded is not what you'd use if you were actually in the situation or because the situation is far-fetched. How often have you used guess-and-check to answer a math problem in real life that in high school would have required algebra? You also might have used algebra in math class to figure out when two trains will speed past each other—and not cared at all. Let them crash; serves them right for not figuring this out before they left their stations.

So, it's not surprising that "math is problem-solving" can sound hollow to someone whose experience with math began and ended in school. Sure, we've all spent lots of time solving math problems. But those problems did not always feel like problems. A problem is not an exercise in using a technique that someone taught us. A problem is not a silly story with artificial urgency that just happens to require the exact mathematical technique explained on the previous page in the textbook. A problem is a question that we care to answer. A problem is an issue that we don't know how to address at first but that we work on because it matters.

A problem is needing to know what the strange symbols on an ancient tablet say. Needing to show that your electoral system is unfair. Needing to get you and your friend out of jail. Needing to know why your peers keep failing a course that they passed last year in a different school. Needing to prove your husband wrong.

Can math solve these problems?

It can. And it does. Mathematicians are often called upon to help solve problems that seem more cultural than mathematical. People tend to think of math as being detached from culture. But it isn't. Math is as much a cultural practice as are anthropology, social science, psychology, and art. People developed math. It arose because people were trying to solve problems that were important to them. The cultural aspects of math are what make it so interesting to people. They also give it power to help solve important social problems.

Sometimes, at least. Other times, math's power blinds those who wield it from its limitations. As Spider-Man knows, with great power comes great responsibility. Improper use of math is itself a problem. A problem that math can also help solve.

"Hi, Amazing Supermath. My colleague thinks his mathematical algorithm shows that we should sequester ourselves, rather than help our neighbors, when an epidemic comes. But I think he hasn't accounted for all of the variables. And that if people take the action he recommends, the epidemic will actually get worse. Can you help?"

"I think I can. Send me your data, and we'll get to work making a better algorithm. And then we'll come up with a mathematical way to check our work . . ."

Acknowledgments

A lot changed over the time it took me to write this book. I moved across the country, finished most of a PhD, wrote two other books, got married, had a baby. The only constants were the wonderful people who helped me as I worked. (And that I was still working on this book, of course.)

First, thank you to Vince Burke at Johns Hopkins University Press for shepherding me through the writing and publishing process. Vince's patience was admirable. Many friends and family read drafts of the book as I went along. Thank you in particular to my parents, Burt and Gerrie Weltman, Phil Revzin, Madeleine Swart, and Chloe Winther. Also thanks to Michelle Wilkerson's Writing Group for taking the time to work on my manuscript, even though it wasn't the kind of thing that people typically bring. It is unlikely that this book would be done at all had it not been for the steadfast support and sometimes painful but always on-point feedback from my husband, Joel Weber. Thank you for not letting me stop. Finally, thank you to my son, Eli, for his timely arrival.

SUPERMATH

1 Is Math the Universal Language?

Math and the Problem of COMMUNICATING ACROSS CULTURES

Dear Aliens: One, Two, Three . . .

In 2003, mathematicians Stephane Dumas and Yvan Dutil sent the message shown below into space. They sent it to five stars as part of a project for SETI, or Search for Extraterrestrial Intelligence. Dumas and Dutil hoped it would be the first of many messages between themselves and alien pen pals.

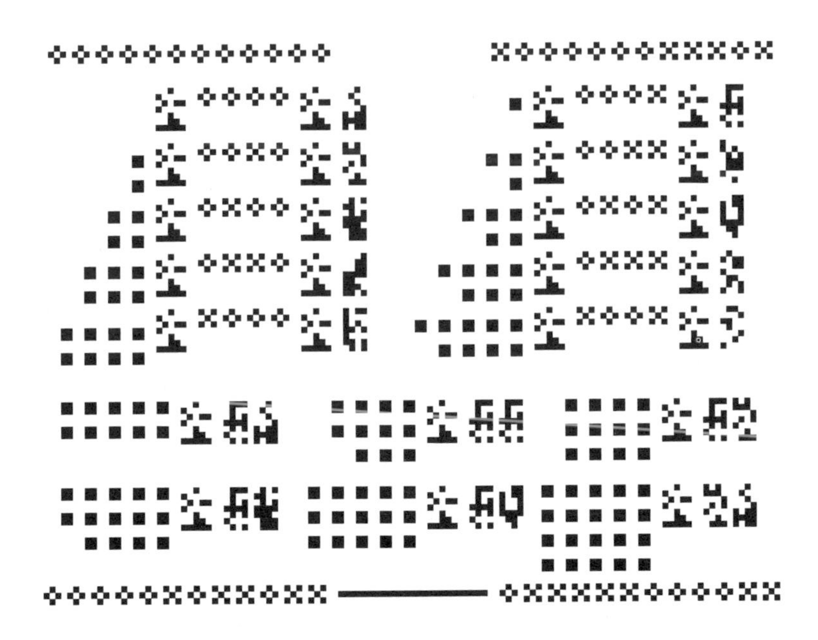

Sending coded messages to aliens may sound like fringe science. But some of the most highly regarded scientists and mathematicians whose sanity has never been questioned have done it. SETI projects have been funded by NASA (National Aeronautics and Space Administration) and the National Science Foundation. SETI's work has led to important scientific breakthroughs, such as improving telescope technology and finding stars similar to our Sun that might harbor life in other solar systems.

But the message shown here looks like nonsense. It doesn't begin in the way we were all taught in school was the polite way to begin a letter. It does not say "Dear Aliens" or identify the senders. And it does not look like any form of written language that we humans typically use to communicate. So, the second most burning question most people ask when seeing this code is, "Why did Dumas and Dutil think aliens would understand it?" (The first most burning question is, "Did they hear back from the aliens?" Answer: No.) You probably don't understand the message yourself. So how could aliens be expected to?

But if you work on it, and I suggest that you try, you'll probably figure it out. Dumas and Dutil thought aliens would, too. That is because the code is math. And not just any math. It's basic counting.

The pixel-like images are arranged to represent our system of counting. But the code doesn't just communicate our basic earthly human system of counting. Subsequent pages of the message show things such as what chemical elements our planet is made of, how many planets are in our solar system, and what people look like. The code concludes with a list of questions that Dumas and Dutil hope the aliens will answer in their return message. The questions are also communicated in pixel-like images. For instance, at the end of the code, Dumas and Dutil ask, "Mickey Mouse ears, attached garage, lobster?," which translates to "Question: Land, extraterrestrials?" Let's hope the aliens aren't offended that Dumas and Dutil think they look like lobsters.

It makes sense that Dumas and Dutil would want to send information about the atomic makeup of Earth and what humans look like to their alien pen pals. All of that is essential to understanding who, what, and where we are. That is, it makes sense if we assume the aliens won't want to use the information to steal our planet from us. Regardless, we can assume our alien pen pals would be as curious to learn these things about us as we are to learn about them.

Strangely, though, this letter doesn't start with our who, what, and where. The first six pages of the message are all math. Basic, abstract math. It's a strange way to start a letter to someone who knows nothing about us.

So, why all the basic math? For a simple reason. And Dumas and Dutil aren't the only correspondents to decide to include counting in their messages. Many like Dumas and Dutil have trusted math to solve one of the universe's most important social problems: How do we communicate with people unlike ourselves?

Why did Dumas and Dutil think that counting could help them communicate with aliens? They did not feel the need to catch extraterrestrials up on third-grade Earth math before they could go on to more important information. They sent math into space for a much more philosophical reason. Dumas and Dutil think that no matter what form space aliens take, if they are intelligent, they will understand math. They think numbers and counting are universal. Numbers, their thinking goes, are the universal language—literally. But are they?

DUMAS AND DUTIL SHARE the common view that numbers and counting form a universally understood mathematical language. They think that, through counting, math gives them the power to communicate with aliens—if the aliens are listening. But is math really the universal language?

Let's examine the code itself. How does Dumas and Dutil's code teach aliens how to count?

The message shows how Earthlings count to twenty in three different ways. You may have noticed that groups of symbols seem to be clumped together. You also may have noticed that some of the symbols share similar features. Each clump of symbols shows a number between one and twenty in three different ways.

First, the message shows numbers using boxes. Dumas and Dutil draw a little box to represent each "one" in a number. If you tried to decipher the code, this is probably the first pattern you noticed. The initial entry has no little boxes in it. The next has one box. The next has two, and so on, increasing by ones up to fifteen, and finally skipping to twenty. A clump of ten boxes represents, literally, "ten."

But that's not all the code says. Each clump of symbols represents an equality. The number shown in boxes is also written in two other common number systems. The symbol after the clump of boxes stands for "equals." After that, the code shows the same number using a row of four symbols. This is the number in binary.

If you tried to decipher the code and got stuck somewhere, it might have been here. Binary is a way of writing numbers using just zeroes and ones. Instead of the symbol for zero, the code uses a diamond arrangement of four tiny squares. Instead of the symbol for one, the code uses a checkerboard arrangement of five tiny squares. Four diamonds in a row stands for "0000," which, in binary, is the number zero. Three diamonds followed by one checkerboard stands for "0001," binary for one. Two diamonds, one checkerboard, and one more diamond is "0010," binary for two. And so on.

After another "equals," we have the third numerical representation. This time the code uses symbols that to me look like pictures drawn in pixels to represent the number in base ten, the system that Earthlings use most frequently to communicate about number. Each digit zero through nine gets its own pixel picture.

Zero is a picture of a house with pixel smoke pouring from its chimney. One is a pixel portrait of a man in a stovepipe hat. Two, a fox-eared robber slinking away with a stolen backpack. That's what I see, at least. Feel free to use your imagination. When the code gets to ten, the man in the stovepipe hat sits to the left of the smoke-spewing house. This stands for "10." Eleven is two stovepipe-hat-wearing men, equivalent to our "11," twelve the man and the thieving fox, and so on.

And so the code teaches aliens three ways Earthlings know how to count to twenty. Dumas and Dutil are not the only people to send messages full of basic math to aliens. In fact, most such messages begin with a tutorial on counting. Dumas and Dutil sent out a similar message in 1999, with almost exactly the same content expressed in a slightly different code. One of SETI's earliest endeavors, the Arecibo message sent in 1974, also begins with the numbers one to ten written in binary. In 2009, scientists Michael Busch and Rachel Reddick attempted to improve on Arecibo and Dumas and Dutil's messages by making a message of their own that included even more math.

Scientists sending messages to aliens got the idea to send basic math into space from early twentieth-century Dutch mathematician Hans Freudenthal. Freudenthal laid the groundwork for the modern theory of extraterrestrial communication in his book *Lincos: Design of a Language for Cosmic Intercourse*. He argued that the biggest challenge we face when trying to communicate with aliens is lack of shared experiences. And the solution to that challenge is to talk to them first about math.

Aliens probably don't know anything about the things we most want to share with them. They may not think about matter as made up of atoms, for instance. They might not describe their location in the universe by proximity to the nearest star. They certainly know nothing about our bodies, religions, greatest cultural accomplishments, and favorite foods. We cannot assume aliens are anything like us. Earth itself holds incredible diversity of life. On this planet alone, single-celled microorganisms hobnob with five-thousand-pound hippos. So, even though we eventually want to talk to aliens about our greatest scientific and cultural achievements, our first extraterrestrial communications, Freudenthal wrote, "cannot have the character of a newsreel." Our messages cannot include information only someone who already knows about Earth would understand.

So, what should we send? Freudenthal argued our first messages to aliens should be about basic math. He claimed math is a language, and it's the only language everyone can understand. Some people might speak English. Some might speak Swahili. Or Thai. Everyone can understand math, Freudenthal argued. While people who speak English or Thai might develop slang that bends rules, math is always the same, no matter where you go. To support this claim, Freudenthal looked at math textbooks from around the world. The textbooks might translate the words for "two" and "square" into the language of their readers. But mathematical expressions such as $2^2 = 4$ and $f(x) = x^2 - 9$ never change. Math is clear, concise, and agreed upon everywhere you go on Earth. Why would the same not be true in outer space?

Freudenthal was far from the first person to think of math as a universal language. Mathematicians in the late nineteenth and early twentieth centuries were captivated by the idea that they could create a universal set of rules that would govern how all mathematical statements were constructed. This "grammar" for math would be built from axioms, or mathematical assumptions that are obvious to everyone. Everything that followed from these universal assumptions would also be universal. Mathematicians Alfred North Whitehead and Bertrand Russell were possibly the most famous pair to take on this challenge. Unfortunately for them, another mathematician proved that their project was impossible. There is no way to make a complete and sensible set of rules for math. But this did not stop Freudenthal from trying to build a universal language with math as its foundation.

Freudenthal called his language Lincos. Sentences in Lincos say both mathematical and nonmathematical things, but they are built entirely from mathematical symbols. Here is something you can say in Lincos:

$10 \in \text{Pri. } 11 \in \text{Pri. } 101 \in \text{Pri. } 111 \in \text{Pri. } 1011 \in \text{Pri. Etc.}$
$1 \notin \text{Pri. } 100 \notin \text{Pri. } 110 \notin \text{Pri. } 1000 \notin \text{Pri. } 1001 \notin \text{Pri. Etc.}$
$\text{a} \in \text{Pri.} \leftrightarrow\text{. a} = 10. \vee.\ \text{a} = 11. \vee.\ 101. \vee.\ 111. \vee.\ \text{Etc.}$
$10 \neq \text{a.} \wedge.\ 10 \text{ Div a.} \rightarrow.\ \text{a} \notin \text{Pri.}$
$11 \neq \text{a.} \wedge.\ 11 \text{ Div a.} \rightarrow.\ \text{a} \notin \text{Pri.}$
$100 \neq \text{a.} \wedge.\ 100 \text{ Div a.} \rightarrow.\ \text{a} \notin \text{Pri.}$
Etc.

The above text means, "Two is a prime. Three is a prime. Five is a prime. Seven is a prime. Eleven is a prime. And so on. One is not a prime. Four is not a prime. Six is not a prime. Eight is not a prime. Nine is not a prime. And so on. A number a is a prime if and only if a is two, three, five, seven, and so on. If the number a is not two but two divides a, then a is not a prime. If the number a is not three but three divides a, then a is not a prime. If the number a is not five, but five divides a, then a is not a prime. And so on."

You might not want to communicate about prime numbers. You might prefer to communicate about more personal things. Freudenthal also has that covered:

Ha Inq Hb. ? x. 10x=101:
Hb Inq Ha. 101/10.
Ha Inq Hb Ben.

This text translates to, "Person A asked Person B, 'What is x if two times x equals five?' Person B said to Person A, 'It's five divided by two.' Person A said to Person B, 'Good.'" There's still math here, but at least this sentence contains some human interaction. Freudenthal figured out how to share praise through math.

That Freudenthal's mathematical language can communicate "good" and "bad" might seem surprising. Prime numbers and solving for unknowns in equations are clearly mathematical. That something could be good or bad is not. Saying something is good is not based in logic, unlike saying something is true. People might disagree about what's true, but someone must be logically correct. Good, however, is subjective. Languages must convey subjective as well as objective ideas. How does Freudenthal's logic-based language handle this?

In Lincos, you can never use the words for good or bad without first saying that a person used those words. This logical rule makes it clear that good and bad aren't properties of objects, but are opinions. "'Ben' and 'Mal,'" Freudenthal writes, using the Lincos words for good and bad, "will be no more than expressions used by some people in order to approve or disapprove of some events." By tying the words "Ben" and "Mal" to individual speakers, Freudenthal invents a logical rule that clearly explains when those words can be used and what meaning they have within a sentence.

Freudenthal defines all of his words using logic. He theorized that a language whose grammar is as uncontroversial as the rules of math has the greatest chance of being understood by aliens. Even if the aliens do not have notions of good and bad, they will be able to follow the logic in the language to learn what we mean.

Many mathematicians since Freudenthal have agreed. Dumas and Dutil did not send their hypothetical alien pen pals a message in Lincos. But they did send a message using grammar aliens can learn by first understanding how we talk about numbers. Like Lincos, Dumas and Dutil's language uses mathematical logic to convey ideas beyond math. Their decision to send math to aliens rests on the premise—or, mathematically speaking, the axiom—

that math is universal. No matter what life is like across the universe, numbers and logic span the distance.

Freudenthal, Dumas, and Dutil must have never visited Papua New Guinea. If they had, they might have met the Oksapmin, an amazing group of people who would not have agreed that math is universal. For until forty years ago, the Oksapmin would not have understood Lincos or Dumas and Dutil's code. They used math that was fundamentally different from what Dumas and Dutil sent into space. And then their math *changed.* Not because someone taught them new math—but because their changing culture compelled it.

Math without Adding

The Oksapmin people live deep in the mountains of Papua New Guinea. They are just one of thousands of indigenous ethnic groups on the island. Linguists have long studied the multitude of languages found there. The Oksapmin recently also caught the attention of mathematicians because of a unique aspect of their language: how they count.

The Oksapmin count using their bodies. You are probably familiar with one form of body counting, using your fingers to count to ten. People of all ages from Western cultures rely on fingers to count and communicate about numbers. Young children learning to count use their fingers to bridge the gap between the words that represent numbers and the number of things they stand for. Even adults who have made that basic connection between numerical words and numbers of things use their fingers to communicate about number. The next time you're buying donuts or asking for a table with friends at a restaurant, pay attention to how often you raise your fingers. "Four donuts" or "Table for four, please" often comes with the gesture of raising four fingers—especially if you are communicating across a wide counter or a noisy room.

In Western culture, fingers are only useful for communicating numbers up to ten. If you wanted a dozen donuts, you probably would not flash ten fingers followed by two fingers. You certainly wouldn't show ten fingers and two toes or, even more weirdly, some other body part that you associate with twelve, like your ear lobe. We operate under a one-more-finger, one-greater-number policy. One ear lobe for twelve would be inconsistent with our way of thinking and communicating about numbers.

But in Oksapmin, touching your right ear and saying the Oksapmin word for ear is how you say "twelve." The Oksapmin use parts of the body to count in ways that stretch the correlation between the number of body parts and number of objects being counted.

Starting with the right thumb, or *tip^na* in the Oksapmin language, the Oksapmin count up to five using the five fingers on the right hand. But here's where their body counting system goes rogue. The right wrist, *dopa*, is six. The right forearm is seven, right elbow eight, and up over the shoulder and across the face. The right ear, *nata*, is twelve, and the right eye, *kina*, is thirteen. The body count reaches its apex at the nose, *aruma*, fourteen. Here the counting system turns around and continues back down the left side of the body. Fifteen is the left eye, which the Oksapmin call the *tan-kina*, with the prefix *tan* indicating that they're on the other side of the body. The count continues to the left ear, over the shoulder, down to the left forearm and wrist, the *tan-dopa*. The count finally travels across the left hand, ending with the left pinky, *tan-t^th^ta*, for a total of twenty-seven.

Once an Oksapmin person has made it all the way from the right thumb across the top of the body to *tan-t^th^ta*, twenty-seven, it's conventional to celebrate. An Oksapmin will raise two fists in the air and shout, "*fu!*" But this is not the end. After the customary "*fu!*" the Oksapmin turn around and go back again, skipping the left fingers, to the left wrist, still *tan-dopa*, for twenty-eight, the left forearm for twenty-nine, and so on.

Body counting such as this was first documented for Western audiences in 1975 by Donald C. Laycock, an Australian linguist who researched the languages of Papua New Guinea. Laycock called the Oksapmin way of counting a tally system. He did this to distinguish the Oksapmin count from "true" number systems, like our Western number system. True number systems, according to Laycock, can be used to add, subtract, and do more complex math. The Oksapmin system, however, can only be used to count.

What Laycock meant by this was that the Oksapmin body count makes it difficult to refer to numbers without counting something. In modern Western culture, we frequently use numbers without counting or referring to actual things. For instance, we often use numbers without counting to do arithmetic. Sometimes the numbers in our arithmetic problems represent quantities of something, but sometimes they don't. Six plus eight is always fourteen. It doesn't matter whether it was six carrots plus eight carrots, six cats plus eight cats, or six piles of five million pennies plus eight piles of five million pennies. Western numbers have lives of their own, independent of the things they sometimes represent. They are what can be called abstractions. They are ideas that can exist and interact only in our minds.

But in the Oksapmin body count, numbers cannot be understood without whatever they are counting. When we hold up four fingers, or point to four objects with our four fingers, the symbol for "four" and the four objects being counted are inseparable. The Oksapmin system may not look like a real tally system because the wrist, forearm, eye, and nose are not additional fingers. Laycock suggested, however, that the non-finger body parts may have entered into the Oksapmin system because each body part corresponded to an additional segment of length when measuring something against the body. Rather than an additional object, each additional body part represents one more unit of length.

Mathematicians call the one-body-part, one-object rule in tally systems the "one-to-one" property. And while the Oksapmin body count and Western fingers count rely on it, Western no-fingers counting ignores it all the time. This difference is huge, particularly because it allows us to easily and efficiently do arithmetic. It is clumsy to do even the simplest arithmetic with the Oksapmin counting system.

For instance, how would you use the Oksapmin body-count system to add six and eight, without having six objects and eight objects to combine?

Here's one approach: starting with the Oksapmin position for one, the thumb, move around the body until you reach the position for six, the wrist. Then, continue moving, to the forearm for seven, inner elbow for eight, and so on, but use the body-part words for one, two, three, and so on. Stop when you say the body-part word for eight. At this point, you will have reached your nose, fourteen. In this way, you can count upward from six to your desired destination, fourteen, while keeping track of how much you've added. Six plus eight is wrist plus the Oksapmin body-count word for eight, which lands you on nose, fourteen.

University of California, Berkeley, mathematics education scholar Geoffrey Saxe calls this method the "body-part substitution strategy." It is similar to the one you might use if asked to add six and eight using your fingers. Young Western children often use it when learning to add. You count to six on your fingers, saying "one, two, three" and so on, and then move on from the thumb that represents six, saying "one, two, three" and so on again. You stop when you say "eight" and reach the finger that represents your second trip past four, for fourteen. This strategy resonates with how Western math works.

Another approach also resonates with Western math. Imagine the body position for six using the right side of your body and the body position for eight using the left side of your body. In this approach, you're not thinking of the left side as continuing the count but as a replica of the right. Then, try to take an amount from one side that when moved to the other side makes ten. Eight is closer to ten than six is, so take enough from six to turn eight into ten. You can do that using the body parts by touching the body part on the right that represents five, one less than six, to the body part on the left that represents nine, and then the body part on the right that represents four to the body part on the left that represents ten. You now have ten on the right and four on the left. If you have a special feeling for ten, you automatically know ten and four make fourteen. This is the answer to the problem.

This approach, which Saxe calls the "halved-body strategy," is a bit harder to visualize. But in many ways it resonates more with Western math than the body-part substitution strategy did. This is because we do have a special feeling for ten. Ask any person who went to school in the United States to add ten to any number, and you are likely to get an immediate response. While the Oksapmin celebrate that they've finished counting when they reach twenty-seven, we might exclaim "*fu!*" upon reaching ten. But instead of saying "*fu!*," what we actually do is record each ten with a new word (ten, twenty, thirty, and so on). All numbers after ten are some stack of tens, arranged into further stacks of tens, with a few ones tagging along.

This base-ten structure that most Western children learn makes working with ten natural for Westerners. Working with ten is not natural in the Oksapmin system, however. The position for ten, the shoulder, is not a privileged position in Oksapmin. The Oksapmin body count isn't even symmetrical around ten. Twenty-seven, the end of the count, or fourteen, the middle of the count, are the most significant numbers in the Oksapmin body count.

If you think about it, the body-part substitution strategy we first looked at as a way of comparing our Western way of counting with that of the Oksapmin also does not feel natural in the Oksapmin body count. Using the body-count words for one, two, three, and so on with the wrong body parts is a perversion of the one-to-one system. You cannot remove the body-part words from their corresponding parts and turn them into objects in their own right to count and add. The body parts do not have lives of their own. They are not abstractions, as are our numbers.

In fact, trying to add eight to six without objects to count in Oksapmin is awkward. That is because the Oksapmin body-count system is not designed to treat numbers as abstractions, as things in themselves. The most reasonable way to add without using objects in the Oksapmin system is to guess. Use the body parts to count to six (wrist). Then keep going. But the Oksapmin have no way of starting to count again, this time to eight, so that they would know exactly when they had reached the body part that represents fourteen. We can do it using our counting system. But they cannot. So if you just keep going in their system without also using our system, how do you know where to stop? Stop when you think you've counted eight more. Did you stop short of fourteen? Did you blaze past it? It is an inefficient way to add without things to count.

Seemingly, the only way for the Oksapmin to add without having things to count would be for them to incorporate aspects of our counting system into theirs. That's what the halved-body and body-part substitution strategies do. But given how unusual the kind of thinking required for the halved-body or body-part substitution strategies is, you wouldn't expect to see any Oksapmin adding in either of these ways. Before the 1960s, your expectations would have been correct. Saxe tried and failed to teach some Oksapmin how to add without guessing and without objects to help their counting. Their understanding of numbers did not permit it. But Oksapmin society changed dramatically in the second half of the twentieth century. And Oksapmin mathematics changed with it.

When Saxe visited the Oksapmin in 1980, he was surprised to find many adults using the body-part substitution and halved-body strategies to add. Plenty of people also guessed, and they got the addition problems wrong. Interestingly, Saxe noticed that those who used strategies that radically altered the counting system were those who'd had the most contact with Western currencies. Something about what the Oksapmin were doing with money had led to a dramatic shift in how they thought about numbers.

Starting in the mid-1960s, some Oksapmin moved to neighboring provinces to work in Australian-run plantations and mines. These workers were paid in cash, first in Australian currency and then, after Papua New Guinea won its independence from Australia in 1975, Papua New Guinea kina and toea. Before this migration, the Oksapmin had never used currency. Their economy worked on a barter system. But once Oksapmin workers began returning to their hometowns with money to spend, the stores they visited had to adapt. Trade store owners had to learn to take money in exchange for the items they sold.

While bartering works with one-to-one counting—three bags of rice for one chicken, for instance—cash does not. Cash comes in denominations that do not cleanly correlate with things bought and sold. Say a bag of rice costs fifteen toea. The buyer could pay for four bags of rice by handing the store owner four groups of fifteen toea, which would be consistent with a one-to-one counting system. But this isn't what typically happens in currency transactions. Toea come in denominations of five, ten, twenty, and fifty. It's easier to exchange four bags of rice for one fifty-toea piece and one ten-toea piece than it is to count out fifteen toea four times. And what if you only had fifty-toea pieces? You would have to pay for the four bags of rice with two fifties, and then the cashier would have to make change, which involves adding and subtracting. Buying and selling with money requires detachment from one-to-one counting.

Not everyone had to learn to use the new currency. Because Oksapmin society in 1980 still did not rely exclusively on currency, many elderly Oksapmin didn't have to learn to work with it. Saxe found the returning laborers and trade store owners were much more likely to use the addition strategies that dramatically shifted the body-count system than were other adults who didn't have to deal with currency.

This correlation between use of the novel addition strategies and money led Saxe to draw a bold conclusion: working with the new currencies had somehow led the Oksapmin to shift their counting system. For adults who used the new methods, the body-count system no longer served a purely counting function. It also served an adding and subtracting function. The words and body parts represented numbers apart from their correspondence with objects. They were abstractions. They had internal meaning, which allowed Oksapmin to use them to do arithmetic without objects. It seemed that those who had encountered currency shifted the fundamentals of their mathematics to solve new problems.

Possibly the most central of those problems was making change. Making change involves doing arithmetic with large numbers, typically numbers larger than those captured by the Oksapmin body count. The Australian and Papua New Guinea currencies were organized around multiples of ten, because both used some multiple of ten of the smaller unit of currency to represent one of the larger unit. These activities increased the importance of multiples of ten and decreased the importance of fourteen and twenty-seven, the focal numbers in the Oksapmin body count.

The Oksapmin also needed to communicate with each other about what they were doing with money. Buying and selling are collaborative activities. Both the person paying and the person getting paid need to agree that the transaction has been conducted properly. Therefore developing a system for keeping track of and communicating about numbers apart from the objects they represent while doing arithmetic with money becomes important. Although the Oksapmin body-count system may have worked to keep track of objects when bartering, it did not work for the more abstract arithmetic needed to use money. To solve these new problems, the Oksapmin needed a new math. And they developed it.

Saxe did more research with the Oksapmin over the next twenty years, during which their society continued to change. Everything he found supported his conclusion: people working together to solve problems can't help but change their culture's mathematics. The changes to counting that Saxe found in 1980 opened a window to help us understand how math developed in the course of history. The story of the Oksapmin shows that not only does math vary greatly across cultures, but also that math changes as culture changes. The mathematics of a group of people both shapes and is shaped by the problems they solve.

So, before 1960, the Oksapmin probably would not have been able to understand Dumas and Dutil's code. Their culture worked perfectly with math that did not involve using numbers apart from things counted. They could have gone on forever using their body-count math if their society had not changed. But it did, and so they changed their math. Math, it seems, is just as cultural as are science or history.

Because the Oksapmin would not have understood the math in Dumas and Dutil's code before the 1960s, does this mean math might not help us communicate with space aliens? Maybe it does, as space aliens might also have a different math culture. Or maybe it means that when we communicate about math, we're communicating about something deeper and more human than logic. We're communicating about our culture. And this is something we want to share with aliens. If the aliens somehow manage to translate our math, they will crack more than just a mathematical code. They will decode a story of who we are as people.

The Oksapmin are an example of a human culture that developed a mathematical system independently and differently than most other humans on Earth. Given their physical isolation from the Eurasian, African, and American continents, it is not surprising that the Oksapmin culture might develop differently. It is in turn not surprising that they and those of us from what has become the mainstream math culture in most of the world might not readily understand each other's math.

But what is surprising is that often we cannot understand what looks like mathematical messages that come from our own ancestors. Archaeologists dig up mysterious messages from the past, messages so strange it's almost as though they are written in code. Sometimes those messages are about math. And it often isn't obvious what they say—even when mathematicians have cracked the code and think they understand the math.

A Trigonometry Food Fight

In the fall of 2017, Australian mathematicians Daniel Mansfield and N. J. Wildberger dropped a bombshell on mathematical historians. "Plimpton 322," they declared, "is Babylonian exact sexagesimal trigonometry."

Plimpton 322 is a Babylonian clay tablet. Clay tablets were the means by which Babylonians kept their records, as they did not have paper or parchment. Scratched onto the fragment known as Plimpton 322 are some symbols that archeologists have struggled to decipher. But now, two mathematicians claimed to have succeeded. And the message was ostensibly a form of trigonometry.

In the midst of the excitement among mathematicians, the mass media attempted to interpret Mansfield and Wildberger's findings for the public. *The Guardian* announced, "Mathematical secrets of ancient tablet unlocked after nearly a century of study." The Australian Broadcasting Corporation claimed that Mansfield and Wildberger's finding would "make studying maths easier." Both articles show the same picture of a grinning Mansfield, holding the famed Plimpton 322 in gloved hands.

Mansfield and Wildberger seemed to have cracked open a mysterious relic of ancient Babylonian mathematics. And it would seemingly shed bright light on the history of math and on the ways that we do math today. But not all mathematicians bought the hype. One mathematician, Evelyn Lamb, called Mansfield and Wildberger's work "outright nonsense." She condemned the publicity video put out by Mansfield and Wildberger's university for peddling in false information.

Mansfield and Wildberger had stirred up some serious drama in the mathematical history community. If you thought that all of trigonometry was inherently dull, think again. This was a trigonometry food fight.

So, what's going on here? What is so controversial about Plimpton 322?

Interestingly, mathematicians do not disagree about what Plimpton 322 says. Everyone agrees about the numbers the symbols on it represent. Plimpton 322 is math. But they do not agree about what it means.

If Mansfield and Wildberger are right, Plimpton 322 is the earliest example of trigonometry and an important forerunner of the trigonometry we use today. It significantly predates the origins of trigonometry from what historians had previously thought. But if they are wrong, it's just an answer key for ancient Babylonian algebra homework. Boring. So, which is it?

PLIMPTON 322 LOOKS UNREMARKABLE. It is an ancient Babylonian clay tablet. Similar clay tablets have been dug up all over the Middle East where the Babylonians once lived. It is covered in little cuneiform marks, the ancient Babylonian form of writing. It was probably made around 1800 BCE in the ancient city of Larsa, in modern-day Iraq. Someone dug it up in the 1920s, after which it made its way to George Arthur Plimpton, for whom the tablet is now named.

When Western scholars first examined the Plimpton 322 tablet, they assumed it was just another ancient Babylonian spreadsheet. The Babylonians ran an efficient empire. They maintained order over a vast territory in large part because of their impeccable recordkeeping. The entries on Plimpton 322 were organized into a table, like many other Babylonian tablets known to hold accounting information. So, this conclusion made sense.

But in the early 1940s, two mathematical historians, Otto Neugebauer and Abraham Sachs, realized some of the entries carefully etched into the pocket-calculator-sized tablet were numbers that formed mathematically interesting patterns. Patterns you would not expect to see in a table of bushels of wheat gathered or livestock butchered. And so began the fight over Plimpton 322.

Some of the numbers looked like what mathematicians call Pythagorean triples. You might remember Pythagorean triples from your high school geometry class. They are numbers that satisfy the equation $a^2 + b^2 = c^2$. This famous equation is named after the ancient Greek mathematician Pythagoras, who supposedly first posed it. It has to do with how the lengths of the sides in a right triangle relate to each other. Squaring and adding together the lengths of the two legs of a right triangle gives you the same number as squaring the right triangle's hypotenuse, the side opposite the right angle.

Plenty of numbers satisfy the equation derived from this theorem. Many are horribly long decimals. But some special numbers are whole numbers, such as three, four, and five, or five, twelve, and thirteen. These are the Pythagorean triples.

Pythagoras lived and worked one thousand years after Plimpton 322 was supposedly created, so mathematical historians were shocked to find Pythagorean triples on it. Were the numbers on the tablet advanced mathematics and not merely a list of things? Was Plimpton 322 evidence that Pythagoras did not create the Pythagorean Theorem?

Even more surprising were the numbers that mathematicians

found in the first column on the tablet. A chunk was missing where the ancient Babylonian scribe would have written the title of the column, so no one could say for sure what the numbers were intended to mean. But if the mathematicians who examined the tablet were correct, these numbers might—just might—be one of the first examples of trigonometry in the history of mathematics. The numbers looked similar to what mathematicians now call *tangent.*

A quick review: tangent is one of the trigonometric functions. The most fundamental tools of trigonometry, these functions deal with relationships between side lengths and angle sizes in right triangles. They are actually simple to construct. The three of these functions with which you are probably most familiar are sine, cosine, and tangent. They are all made by picking one of the two non-right angles in a right triangle and taking ratios of the side lengths around it. You can find the tangent of an angle, for instance, by taking the ratio of the length of the side opposite to it and the length of the side adjacent to it.

Because the trigonometric functions are calculated by finding a ratio, the tangent of an angle is always the same, no matter how big or small the right triangle in which it sits. This makes the trigonometric functions very powerful. If you know the exact trigonometric values of any angle, you can learn almost anything you want to know about a right triangle.

Knowing the exact trigonometric value of any angle should not be taken for granted, however. On a calculator, the trigonometric functions turn measures of angles into never-ending decimals. A calculator somehow knows that the tangent of forty-two degrees is 0.90040404429, and so on. The precise calculation of tangent generally results in decimal numbers so long that it would be enormously time-consuming for a person to calculate them by hand. You have to have a calculator to work with trigonometric functions. How the calculator knows this, however, is a mystery to most people who use calculators. Calculating the trigonometric values of most angles requires some pretty advanced mathematical techniques.

Remarkably, it looked like whoever wrote Plimpton 322 might have known how to calculate tangent, or something similar to it, without having to use messy decimals. That would have been a mathematical miracle.

The numbers on Plimpton 322 that conjure up ideas of tangent are organized so as to increase in value as you move down the table. A modern trigonometric table of numbers would be organized in this same way. But values of tangent on a modern trigonometric table would correspond to angle increases of one degree. Our tables show the tangent of one degree, two degrees, three degrees, all the way up to three hundred and sixty degrees, at which point the table starts again at the beginning.

But the numbers on Plimpton 322 do not seem to ascend in sequential steps corresponding to angle increases of one degree. Plimpton 322 seems to ignore angles altogether. Remember, trigonometric values of most angles are messy. They are long and complicated and can only be calculated efficiently with something like a calculator. They are what mathematicians call irrational numbers, numbers whose decimals never end and cannot be written as fractions. For this reason, you can never write all of the numbers that make them up. Mathematicians decide to produce trigonometric values with so many decimal places and then say, "That's good enough."

Yet it is possible to compute a tangent without angles, using only side lengths. And if you do that, you can avoid irrational numbers altogether. You won't need a calculator. All of your trigonometric values become tidy fractions, because you made them that way. You can choose to use side lengths that result in simple trigonometric ratios. For instance, if your triangle has side lengths three, four, and five, then the trigonometric ratios you create will be three-fifths, four-thirds, and other pleasant fractions involving those simple numbers. Never mind the sizes of the angles in that triangle, for the side lengths are enough. According to Mansfield and Wildberger, this is how tangent is calculated in Plimpton 322. This is where their claim that Plimpton 322 holds "exact" trigonometry comes from. Trigonometric tables that are constructed from angles require an enormous amount of computing. And they can never be exact because there isn't enough room on a table to write an infinitely long decimal. But trigonometric tables built without angles, just with triangle side lengths, always are exact.

So, if you believe Neugebauer, Sachs, Mansfield, and Wildberger, Plimpton 322 is a historical bombshell. The ancient Babylonian who wrote Plimpton 322 beat the Greeks to the Pythagorean Theorem by a thousand years *and* did impeccably precise trigonometry. Plimpton 322 is "the world's only *completely accurate* trigonometric table," Mansfield and Wildberger claimed. It is time, they argued, to rewrite the history of mathematics, and along the way reevaluate the status of our culture compared with that of the ancient Babylonians.

Exciting, right? Not so fast, say other mathematicians. Let's not knock Pythagoras off his pedestal just yet. Remember, Evelyn Lamb called Mansfield and Wildberger's claims nonsense. So, what's the opposing argument?

The claim that Plimpton 322 contains trigonometry conflicts with the conventional historical narrative about trigonometry. But it also corresponds to some popular presumptions about the development of math. In fact, the conclusions that Plimpton 322 contains trigonometry support a popular narrative about math.

Mathematical historians often seek to trace the development of modern Western mathematics in the works of revered ancient cultures. There is a tendency to think that if a culture was great and produced wonderful works of art, architecture, and science, it must be like ours. The Babylonians are one such culture. Historians often talk about the Babylonians and other groups that settled in the Middle East as the seeds from which modern Western civilization bloomed. The pre-Islamic Middle East was where, as the story goes, "man makes the giant step from savagery to civilization." This story leads to faulty and xenophobic logic. In this view, people who don't think like we do today were savages. People who think like we do are civilized. Ergo, since the Babylonians were clearly civilized, they must have thought like we do. They should have been able to develop trigonometry.

The conclusions of Neugebauer, Sachs, Mansfield, and Wildberger that Plimpton 322 holds trigonometry feed directly into this narrative. But when interpreting documents from long-lost cultures, we must be careful to keep an open mind. We must not let our predisposition to romantic stories get in the way of the facts. So argues Eleanor Robson, a scholar of ancient Middle Eastern history at University College, London. Robson accuses these scholars of fitting the tablet like a puzzle piece into a narrative they have already constructed. She argues they are behaving like Hercule Poirot in a classic British murder mystery. In "The Mystery of the Cuneiform Tablet," as she calls it, the scholar, a Poirot-esque sleuth, already knows who committed the crime. He just has to convince the local cops. So, he uses bits of compelling evidence at hand to craft a tale he wants to be true. Plimpton 322 is a convenient piece of evidence for telling a story about mathematical history that lots of people want to be true but might not be.

Interpreting ancient artifacts is never tidy work. Mathematical artifacts were made by real people, the products of messy and ever-changing cultures. Artifacts buried for centuries outlast the intangible struggles, goals, and dreams of the people who made them. Without information about those struggles, goals, and dreams, scholars cannot draw robust conclusions about the objects ancient people left behind. Scholars are left guessing like kids playing a game of Clue, rather than deducing like the genius Poirot. Is it Babylonian mathematician, inventing trigonometry, with the clay tablet? Or Babylonian algebra teacher, checking homework, with the clay tablet? All we know for sure is that a clay tablet was involved.

Robson argues that interpreting mathematical texts from cultures different from our own must be done with the culture that created them in mind. It's useful to know what the Babylonians were doing with Plimpton 322. What Plimpton 322 scholars are missing, according to Robson, is that what we mean by "trigonometric table" today would not have made sense to ancient Babylonian mathematicians. Even though we use the same number patterns and do the same calculations, Babylonian mathematics was fundamentally different. To fully understand the numbers on Plimpton 322, we have to look more closely at what problems the Babylonians would have used them to solve.

And, Robson argues, they probably weren't doing trigonometry. That's because, according to Robson, the ancient Babylonians didn't use angles. Trigonometry is all about angles. Even if you calculate trigonometric functions using the ratio method that Mansfield and Wildberger say is in Plimpton 322, you can't entirely escape angles. The power of trigonometry lies in being able to attach ratios to angles in right triangles. Otherwise, there is no reason to do it. It would be a useless exercise.

But the Babylonians didn't have a concept of angle. They never measured angles or even talked about them. It's not that they didn't see an angle when it was in front of them. It's just that angles did not figure into anything they were doing. And with no angles, it's unlikely the Babylonians were doing trigonometry.

That the Babylonians could have been doing any math without thinking about angles might sound surprising. But an angle is a tricky mathematical object. Try to come up with a good definition for "angle." You'll have a hard time. You can't just say vague things like, "You know, it's the pointy part of a triangle," and wave your hands around. Angles are hard to think about because they are intangible. They are purely abstract. Lengths, areas, volumes, and other mathematical objects that we measure can be picked up to determine their weight or measured against rulers. These are discrete, tangible things. But you can't do that with an angle. You can have the same thirty-degree angle in a tiny triangle and in a giant triangle. Angles feel as if they get wider as you move away from the pointy part. And the pointy part feels as if it doesn't have any size at all. Because of this, angles are more abstract than other geometric objects that we take for granted. It makes sense that people might do math but not think of an angle as something worth measuring.

Robson does not deny that Plimpton 322 is covered in numbers that, to us, look like trigonometric values and Pythagorean triples. She questions, however, whether mathematicians of ancient Babylon would have used those numbers in ways we would expect from a modern perspective—that is, as trigonometry. Her hypothesis is that Plimpton 322 was a sort of answer key used to check algebra homework. Scholars know that the ancient Babylonians went to school. The methods they used for solving algebra problems would have required the sort of numbers written on Plimpton 322. There's no evidence, however, that they did trigonometry.

Robson's story about Plimpton 322 is much less exciting than one in which the clay tablet shatters assumptions about the history of math. And it is hard to know which to believe, because we do not know for sure what problems the Babylonians were solving with Plimpton 322. The debate surrounding Plimpton 322 revolves around whether two things that look alike serve the same purpose. The math on the tablet may look like trigonometry. But if the Babylonians had no use for trigonometry, then maybe the math on the tablet is really something else.

So, in order to really know what the numbers on the tablet mean, you have to do more than try to decode the math. You have to know about Babylonian culture. You have to know what kind of math that culture needed to solve its most important problems. Some of the problems the Babylonians were solving were similar to ones we solve today. But a lot of them were quite different. We face problems that the Babylonians did not have and use math such as trigonometry to help solve them.

Different problems lead to different approaches to even the most basic elements of math. Shapes and patterns that we assume have fixed mathematical meaning, such as angles and Pythagorean triples, may have meant something different to mathematicians thousands of years ago.

The Mystery of the Knotted Strings

Officials in the remote Andean village of San Juan de Collata had guarded the documents for hundreds of years. They had never shown them to an outsider, especially not one from Europe. But perhaps now they'd found someone they could trust. Someone who could tell them what the documents said.

"If we could read what is in here," a village elder told that special person, "we would know for the first time who we truly are." For, while the themes of the documents had been passed down in oral lore for hundreds of years, no one alive could confirm what the documents really said. Because no one knew how to read them.

Maybe anthropologist Sabine Hyland could help.

Hyland isn't a mathematician. But her visit to San Juan de Collata and her encounter with the treasured documents drove the last nail in the coffin of a theory about the Inca that had been developed by a mathematician over a hundred years earlier.

Mathematicians thought these documents, called *khipu*, were records of Incan mathematics. Most believed these khipu held no more than numbers and simple arithmetic. But Hyland and anthropologists like her were about to show that many of them weren't just that. In fact, khipu might be stories. Stories with characters, drama, and deep historical meaning to the Inca who created them long ago.

How did mathematicians get khipu so wrong? Surely, it cannot be difficult to distinguish between arithmetic and a story, no matter how foreign the language in which it is written. But, as the debate over Plimpton 322 also shows, interpreting ancient mathematical records is much harder than you might expect. What looks like just math to us might be more than just math to other people. Math differs from society to society as much as do language and culture. Mathematicians have long been tempted to see our mathematics in the math of other cultures, even when it isn't there. That is, we tend to see our ways as the best ways and value other cultures insofar as they are similar to ours. And we tend to see other people's pasts as merely the prologue to our present.

But that temptation can lead us astray. Sometimes that temptation leads to academic debates over the significance of a clay tablet. Sometimes it leads to something much worse: the denigration of another people's culture and the erasure of hundreds of years of that people's history.

KHIPU, THE TYPE OF DOCUMENT Hyland works with, have long baffled both Incan descendants and anthropologists. For one thing, khipu look strange, as far as documents go. At least Plimpton 322 is readily identifiable as a document, with its obvious writing and table structure. Khipu, however, consist of strands of knotted string. No other culture outside of the Inca has created anything like them. But khipu seem to have been common in the Incan world. Almost everywhere archaeologists find remnants of Incan civilization, they also find large stashes of knotted strings.

High in the mountains of Peru lies the ancient Incan site of Puruchuco. Puruchuco was once a major administrative center of the Incan Empire. Gourds and shells, corn and beans, textiles, ceramics, and metal—all items you might expect to find in an ancient city—were sifted from the five-hundred-year-old rubble. None of these items was particularly surprising, from our modern perspective. We expect ancient cultures to have had many of the things that we do today. But the archaeologists also found an urn full of intricately knotted string. Not rope, not the beginnings of clothing. Simply string, knotted in a variety of patterns.

Khipu consist of a main backbone cord. From that backbone, many more cords made of thinner string dangle. These thinner strings are tied in knots. Sometimes the strings are colored, solid, or in stripes. Sometimes the thin strings carry yet more strings, branching off like a snake with many heads.

When tangled, khipu look more like refuse from a knitting project gone awry than anything useful. But when displayed with care, showcasing the interplay of colors, lengths, and knots, khipu are striking. If you look closely at a khipu, you might notice that the knots are tied in patterns. These patterns provoke a thought: these must have been made to communicate something.

But what?

Khipu tempted anthropologists looking to answer a perplexing question of Incan history: Did the Inca write? The Incan Empire was one of the most organized and advanced of its time. Most other societies like it at some point developed a system of writing. The Babylonians had cuneiform, for instance. Writing is useful for communicating across the great expanses of an empire and recording a civilization's discoveries for people of the future to build on. But anthropologists had never found any remnants of Incan writing. They did, however, find khipu, those bizarre strands of knotted string.

In 1912, mathematician L. Leland Locke claimed to have cracked the khipu code. Khipu, he wrote, were math. He claimed that the patterns found in khipu represented numbers written in a base-ten system much like our own. The Inca had knots for ones, tens, hundreds, and thousands. Locke claimed khipu contained ten different types of knots that carried intentional meaning, all of it numerical. Now, not everything in the khipu seemed to have the numerical meanings he identified. Locke just disregarded these things. Anything in khipu that didn't follow the number pattern he had identified he considered decorative and irrelevant to the meaning of the encoded mathematical message. While the Inca actually may not have written words, Locke concluded, they did write math.

Locke's claims were persuasive to others like him who valued mathematical arguments but knew little about life in the Incan Empire. His conclusions required little context to verify. He could justify them simply by identifying patterns. Locke did just that. He did not do extensive research into Incan culture. It was hard to refute interpretations of these Incan texts that drew on mathematics seemingly familiar to modern-day Westerners without examining how the khipu might fit in with a deeper understanding of Incan culture. Within the academic communities where Locke and his ideas circulated, drawing conclusions about khipu without consulting Incan history and memory was acceptable.

This was partly because little of Incan culture survived the Spanish conquest. It was hard to find a cultural context within which to fit the khipu. The Spaniards destroyed almost everything of value to the Inca as they swept across the Andes. As Cieza de Leon, a Spanish soldier who participated in the Spanish conquest, wrote, "Wherever the Spanish have passed, conquering and discovering, it is as though a fire had gone, destroying everything it passed." Khipu were a special target of Spanish destruction. The Catholic Church claimed khipu were the work of the devil and ordered all khipu burned. By Locke's time, only around four hundred khipu remained. Only four hundred examples of an artifact that seems to have served an essential function in Incan culture.

But not all was lost. Some information about khipu was available, if only Locke and other researchers were willing to use it. This information largely disproved Locke's conclusions. Chronicles from the time of the Spanish conquest told of khipu that held stories. A Jesuit missionary, for instance, wrote about an Incan woman who brought a khipu to confession. She claimed it told her life story. A group of Spaniards in the early 1600s came across a man carrying khipu that he said recorded all the Spanish had done in the Incan Empire since they had arrived. Unless the Incan woman's life story and the man's history were all arithmetic problems, the khipu they carried must have recorded more than just numbers.

Locke's conclusion that khipu were math, no more and no less, stood for almost a hundred years. We can try to understand the challenge that Locke faced in translating khipu. With Incan cultural memory fading, archaeologists and anthropologists found it difficult to make sense of khipu. This difficulty should not be understated. Even descendants of the Inca did not know how to read khipu. Furthermore, Locke and his colleagues had no cultural overlap to draw on when deciphering khipu. In contrast, archaeologists uncovering the walls of the buildings at Puruchuco had no trouble identifying them as walls and quickly discerning their purpose, because the archaeologists were familiar with walls. But Locke had nothing to relate to khipu. Nothing, that is, except his knowledge of base-ten counting and Western mathematics.

But even understandable cultural ignorance does not seem like a good excuse for some of the other shenanigans Locke pulled when writing about khipu. Locke ignored evidence right under his nose that suggested his conclusions were incomplete. For instance, roughly a third of the khipu that archaeologists found did not follow Locke's rules. That's a lot of evidence to ignore. By all accounts, Locke was mathematically perceptive. But he was not a very good historian. He disregarded spotty historical records and rationalized evidence that diverged from what he believed.

How did mathematicians and anthropologists eventually prove Locke wrong? They had to rely on more than their mathematical prowess. In fact, they had to combine their knowledge of mathematics with their knowledge of culture.

As it turned out, the details that Locke wrote off as decorative were the keys to a much more complex system of communication than just the simple arithmetic that Locke had discovered. Eighty years after Locke, mathematician Marcia Ascher and her anthropologist husband Robert figured this out. The key was both deeper mathematical understanding and better understanding of Incan life. The Aschers did something that Locke never did: they thought about *why* the Inca might have needed the mathematics in the khipu.

If khipu were math, the Aschers thought they must have been made to solve some sort of problem. That's often what math is for—solving problems. Knowing what problems the Inca were solving with the khipu was critical to translating them.

Questions relating to commercial transactions and population control would be central to the Incans. How many bushels of corn should we send from Puruchuco to another Incan city to make up for the slow harvest there? What's the best way to keep track of census information we've collected so that we can see changes in our population over time? The solutions to these problems that khipukamayuk—the Incan name for khipu makers—at Puruchuco and elsewhere in the Incan Empire were working on certainly involved numbers. But what did those numbers represent? And how did the Incans convey those representations to each other?

The Aschers thought about how we, today, keep track of commerce and census data. Our records are never simply lists of numbers. Rather, we include information about the purpose of the numbers. We label them. And we structure our records according to our problem-solving needs. To the Aschers, then, it seemed unlikely that the Inca would have failed to include the same essential information in their khipu.

There are many different ways to organize information. A list, for instance, is good for keeping track of what you've purchased from a store, as on a sales receipt. But lists are not the best ways of organizing all types of information. Would a store manager use a list to keep track of all the items the store carries? Maybe. But maybe a table would be better. With a table, you can record several pieces of information at once. You can quickly find information under several different categories. Tables are useful for, say, recording all of the different types of clothing the store carries (shirts, pants, dresses, socks, etc.) and the brand of each item, and then also all of the sizes and prices of each item.

Then again, a table might not be the best way to organize information if a store manager wanted to track how sales of such-and-such-brand shoes compare with sales of this-and-that-brand shoes over time. The manager might instead use a graph. While all information could be captured in a table or on a list, a graph could include the information and also provide a time line. Without a time line, trends and the passage of time would be harder to record. A graph may be harder to construct, but it provides a more complex overview of information.

Lists, tables, and graphs and how stores use them to record information may seem like the stuff of middle school math, not ground-breaking mathematical history. But the ways we structure numerical information have a huge impact on how we make sense of it. And vice versa.

In a complex society like the Incan Empire, it would be surprising if khipukamayuk weren't driven by the need to develop complex ways of organizing and communicating important information. The forms of their numerical representations must have served functions particular to their society. The ways in which they assembled and transmitted information on the khipu may have been different than the ways we do today. But that doesn't mean their methods weren't sophisticated and effective. If we have not been able to fully decipher the khipu, that may be because we have not fully understood Incan society.

This is what the Aschers tried to do. With the additional cultural information that Marcia and Robert Ascher gleaned from thinking about why the Inca made khipu, they found Locke's conclusion to be partially accurate but incomplete. Khipu must record more than just numbers. Learning more about the social problems the Inca were solving with math helped the Aschers interpret the patterns Locke had written off as insignificant decorative flourishes.

For instance, the Aschers noticed that khipu makers used color, grouping, and layering to organize the numbers recorded in khipu into categories. While they did not know precisely what those categories were, the Aschers could recognize when colors and structure were used to add numbers across several categories. They could also see when colors and structure were used to organize numbers into a table. Yes, the Inca created tables using knots. Where Locke had simply seen naked numbers, the Aschers saw a complex system of record-keeping. That system helped the Inca maintain and grow their civilization, much like writing would.

Rather than starting, as Locke did, with our numerical system and then looking for ways in which the Incan system matched it, the Aschers started with Incan culture. Then they looked for ways in which the khipu might represent a numerical system that helped to solve the Incans' social problems. The Aschers' use of culture as an important context for interpreting khipu cracked open a field of research that had once been closed. If the colors and structure in khipu served as organizational markers and not just decorations, then khipu scholars thought it possible that the khipu held much more than was previously thought.

For instance, wouldn't it have been useful to the Incans for khipu to be labeled? There might be things in the khipu that serve as labels. Maybe the items that khipu counted were described somewhere in the document. There might be explanations of the numbers. Perhaps khipu that served as messages from one administrative center to another would have a signature at the end, so that the receiving party would know who sent it. The Aschers' discoveries led to a flurry of research. Scholars went looking for what before would have seemed impossible to find: words.

And almost twenty years later, they deciphered their first word in the knots. In the khipu found at Puruchuco, Gary Urton and Carrie Brezine found a knot pattern that didn't look like any of the numerical knots they had already translated. Locke would have ignored this pattern because it didn't fit his existing schema. Inspired by the Aschers, however, Urton and Brezine went to work translating it. They think this knot pattern says "Puruchuco."

Urton and Brezine's finding verified what Incan descendants had claimed for hundreds of years. Khipu hold words as well as numbers. Locke's conclusions had swept the Inca's historical memory out of popular view. Maybe now scholars with the tools necessary to translate their long-lost records would be willing to listen.

Perhaps this is what motivated the Incan elders of the town of San Juan de Collata to finally share their khipu with Sabine Hyland. The residents of San Juan de Collata claimed that their khipu told the story of an Incan rebellion against Spanish colonial authorities. There certainly might be some numbers in this story, but the bulk of it wasn't math.

The residents of Collata took a risk in sharing their khipu with Hyland. What if she dismissed their claims, as Locke and other scholars had for a century? But Hyland worked with them. Hyland may be an expert in using patterns to decipher mysterious texts, much like Locke and other mathematicians. But the residents of Collata are experts in something just as important for translating khipu: Incan culture.

Most of the information recorded in these khipu remains a mystery. But analysis done by Hyland and the Collata elders has shown that the Collata khipu contain far more variety in color, knotting, and structure than other previously studied khipu. The Collata khipu contain ninety-five different symbols. That's enough to be a system of writing in which different symbols represent different syllables in words.

Hyland claimed that these khipu did in fact contain records in a system of writing called logosyllabic. She and the Collata elders translated a few groups of strings on the khipu. The last group of strings on the first khipu, Hyland claimed, spells *Alluka*, a Collata family name. The last group of strings on the second khipu spells *Yakapar*, a family name common in a nearby village. They were, Hyland hypothesized, the signatures of the people who had made these khipu hundreds of years ago. They weren't math; they were names.

In the Puruchuco and Collata khipu, Hyland, Urton, Brezine, and the Collata elders found some of the first evidence that khipu didn't just record numerical information, as Locke had claimed. Other khipu with patterns as unusual as those found on the Collata khipu exist. In fact, many of these khipu were called forgeries when they were discovered because of how different they were from khipu known to hold math.

Now that anthropologists and mathematicians know khipu can contain more than math, perhaps they can return to those previously rejected khipu with new perspective. Locke's seemingly innocuous mathematical conclusion may have swept significant parts of Incan culture out of view. But now, with their powers combined, mathematicians and cultural experts can decode even the most mysterious texts. Troves of Incan history await.

2 Can Math Predict the Next Move?

Math and the Problem of **WINNING** (or Not Losing, at Least)

A Mathematically Impossible Peace

The depths of winter, 1914. Europe was six months into World War I, one of the bloodiest wars the world has ever seen. Soldiers were dug deep into their infamous trenches. Covered in mud, they picked each other off across no-man's-land. Poke your head above the dirt barricades, whether out of recklessness, heroism, or simple boredom, and you risked death. Death in the form of a fast-flying bullet, fired by someone too far away to see.

And yet, on Christmas Eve of 1914, the shooting stopped. Soldiers up and down the Western Front began calling Christmas greetings to one another. Then singing Christmas carols. Then putting up Christmas trees. And, finally, emerging from their trenches to shake hands. The scene was almost jovial. One British soldier wrote in his diary that a member of his group met a German patrol who gave him "a glass of whisky and some cigars, and a message was sent back saying that if we didn't fire at them, they would not fire at us."

How did this happen? It's hard to imagine what would lead two huge armies to spontaneously call a truce. Generals did not negotiate it. Individual soldiers, for some reason, ignored their rational feelings of fear and stepped out of the trenches to enjoy Christmas with their enemies. Who was that first crazy soldier who risked his life to share some Christmas spirit? And who was that other crazy soldier on the opposite side who didn't shoot him?

The Christmas Truce of 1914 has gone down as one of the most baffling events in the history of warfare. It makes little sense on a gut level. It also makes little sense from a mathematical perspective. Yes, math has something to say about the Christmas Truce. Mathematicians use math to understand how people behave during conflicts. And the rules of math say that a truce like this shouldn't happen.

Tic-Tac-Toe and Other Games Math Has Ruined

There's an entire field of math devoted to predicting human behavior during conflicts. This field is called game theory. Mathematicians who study game theory try to come up with logical explanations for how people will interact in competitive situations, or games. Then they figure out what people should do in those situations to guarantee the best outcome. This is called "solving" the game. Once a game has been solved, you don't have to think to play it. Follow the mathematically determined rules, and you are guaranteed victory. Or at least you are guaranteed not to lose.

You don't have to be a professional game theorist to solve a game. In fact, you've probably solved at least one yourself.

Do you remember when tic-tac-toe stopped being fun? For me, it was during the spring of the year I turned twelve. I was tasked with running a carnival game for preschoolers. I remembered happy times in my younger days playing tic-tac-toe with friends on a chalkboard in my basement. So, I drew nine squares on a portable chalkboard and, with my basket of cheap prizes, sat down to play. I assumed that most children knew how to play tic-tac-toe. Sometimes they might lose, so my prizes wouldn't disappear too quickly. But sometimes they might win. The kids would have a good time. Tic-tac-toe is fun, right?

After the fifth sad child left my station without a prize, I realized I had miscalculated. I played game after game of tic-tac-toe with four- and five-year-olds. Each time I won. Occasionally we tied. But mostly I won.

I played basically the same way each round. This was my strategy: if a coin-flip determined that I went first, I made sure to secure a corner spot.

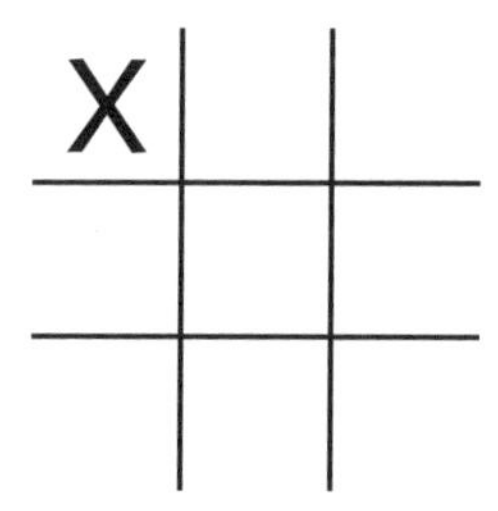

On my next turn, no matter where my opponent played, I snagged another corner. This set me up for three in a row if I could only take the square in between.

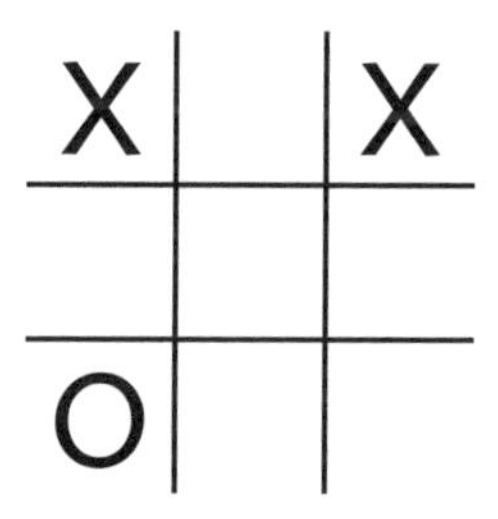

Even five-year-olds could usually block that obvious move. But it didn't matter. Because on my third turn, I took yet another corner. A third corner would always be available. My opponent could only have taken one corner before having to play a defensive move on her second turn.

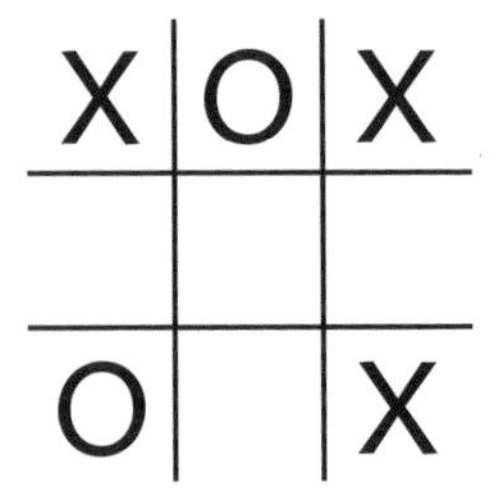

At this point, the game was basically over. Follow this strategy to its inevitable conclusion yourself to see what happens. With three corners in my pocket, I was always in one of two excellent positions. I could either snag three in a row in two different ways on my next turn, only one of which the poor child facing me could prevent. Or our game would end in a draw. My path to either certain victory or to a draw, and a teary prizeless preschooler, is clear in the sample game I've been playing here.

If the coin-flip forced me to start second, I took the middle square. I followed my middle-square opening move with a middle-edge square grab, landing us here:

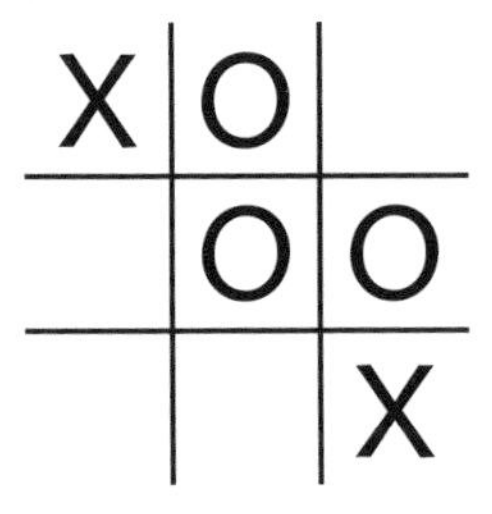

This move forced the first player to block my intended three in a row. From here, the game moved predictably toward a tie, as each player defended against the other's victory.

My day spent clobbering children at tic-tac-toe taught me two things. First, sometimes it was okay to throw a game if it meant that fewer children fled in tears. Second, tic-tac-toe is not fun. The end of the game was fixed. I could always win at best, and tie at worst. What fun is there in a game with a certain ending?

I hope that your experience discovering that tic-tac-toe is a terrible game involved fewer unhappy children than mine did. But at some point, each one of us makes the transition from tic-tac-toe-loving child to tic-tac-toe-jaded adult. It is a rite of passage. If I asked you to give a title to a category of games like tic-tac-toe, you might call them "boring games." Mathematicians have another name for them. They call tic-tac-toe and other games like it "solved games."

A solved game is a game that mathematicians have figured out how to win. Or at least how not to lose. Most adults have solved tic-tac-toe. How I played the game against those poor preschoolers is a form of the solution to the game. If two players both use the solution against each other, the game ends in a tie. If one player uses it and the other does not, the player using it wins.

Solving tic-tac-toe is rather easy. It also ruins the game. But game theory doesn't always ruin games. Sometimes the game actually can't be solved. Other times the solution is so complicated that no one can implement it. And still other times, the game has such high stakes that we desperately need a solution.

LET'S EXAMINE THE FIRST SITUATION of the unsolvable game. Not all games can be solved. But plenty of people would have you believe otherwise. That's because they want your money.

"By this time tomorrow, one lucky person could be $363 million richer! So how do you pick that winning ticket? Well, turns out it's a combination of luck and strategy," said a bouncy *Good Morning America* reporter to an eager Robin Roberts. They were talking about the Mega Millions Jackpot, which on that spring morning in 2012 had just hit a record. Three hundred sixty-three million dollars is a whole lot of money—more than either Elton John or

Britney Spears possessed, according to the reporter. She predicted we would all run to our local convenience stores to try to snag the winning ticket.

But for the jackpot to be that huge, many people must have already bought tickets and lost. Our chances of doing the same were large. As the news story flashed to a video of Britney Spears performing her hit song "Lucky," conveniently fading out just as Britney sang the lyric "But she cry, cry, cries in her lonely heart," the reporter asked, "How can we increase our odds of winning?" In other words, how can we *solve* the lottery?

Lucky for us, we don't have to come up with a solution to the lottery all by ourselves. Richard Lustig claims that he already has. And he generously shared his solution in this television report as well as in his transparently titled book, *Learn How to Increase Your Chances of Winning the Lottery*. Lustig claims to have a "mathematical" strategy for beating the odds and winning the lottery. Except that, unlike the solution to tic-tac-toe, Lustig's strategy doesn't work.

Lustig is just one of many logic-wielding experts who claim to have used math to crack the code of chance and see through the haze of probability. But have these people really found a mathematical way to beat the odds? Is Lustig a mathematician, or just a mathemagician? The best defense against mathemagic is math. So let's use math to examine Lustig's claims.

Lustig advises us not to buy too many tickets. He argues that our odds of winning do not increase just because we have more tickets. This statement is effectively true. Your chances of winning the lottery are so microscopic that buying ten tickets instead of one does practically nothing to increase them.

But this statement is mathematically false. To examine why, let's simplify the situation by pretending that the lottery consists of someone rolling a six-sided die. Your odds of winning if you buy one ticket will be one in six, much higher than they are in the actual lottery. If you buy four tickets instead of one, your odds of winning will increase dramatically to four out of six. Then, if any of the four numbers on your tickets comes up, you'll win. Four out of six is much better odds than one out of six.

But the lottery is not someone rolling a die. The machines that spit out lottery numbers are much more complicated than that. Your odds of winning the lottery if you buy just one ticket are far smaller than one in six. In fact, they are about one in several hundred million. If you buy ten tickets, your chances are ten in several hundred million. Not much difference. For a better-than-even chance of winning the Mega Millions Jackpot, you'd have had to buy three hundred tickets a week since the last Ice Age. So, while adding a few more tickets to your current purchase increases those odds, it does so only by the tiniest amount. The advice not to buy a lot of tickets is solid, even if the math behind it is faulty. But no harm done.

Unfortunately, Lustig also gives advice that is both mathematically flawed and unhelpful. About halfway through the television report, Lustig leaned into the camera and chastised the viewer, "Do not play quick picks!"

"Quick picks" is lottery lingo for the numbers the ticket machine gives you if you allow it to choose your numbers. There are many reasons for not letting the ticket machine choose your numbers, most of them superstitious rather than mathematical. Maybe you think you have a lucky number. Maybe you like the numbers you get from fortune cookies. Or maybe you don't trust the ticket machines to be truly random.

But this isn't why Lustig doesn't want you to play quick picks. He has a mathematical (or is it mathemagical?) reason.

Lustig wants you to use what lottery strategists call the "low-frequency strategy" for choosing lottery numbers. The argument here is that you have a better chance of winning if you pick numbers that have not come up on winning tickets. Winning lottery numbers are chosen at random, so all numbers have an equal probability of eventually being winners. Proponents of this strategy argue that numbers that haven't recently appeared on winning tickets are more likely to do so than numbers that have come up before. This supposedly "evens out" the frequency that different numbers are winners.

Using this "low-frequency strategy," the reporter doing the story for *Good Morning America* recommended that everyone buy tickets with the numbers 12, 13, 41, 55, and 56. Apparently these numbers had not appeared on a winning Mega Millions ticket in twenty-five years. Robin Roberts eagerly wrote them down. But should you?

For starters, if everyone watching *Good Morning America* won the jackpot with those numbers, they would have to split the money almost five million ways. Each winner would get about seventy dollars. So you have a better chance of winning a lot of money if you choose different numbers.

But does choosing numbers that have not recently shown up on winning tickets increase your odds of winning? Do the frequencies of winning numbers really even out over time?

No. The rules of probability say that each number has the same probability of appearing each time the lottery ball machines spit out a ball. Frequency doesn't matter. The balls know nothing about which numbers won in the past. All they know is that the machine is on and balls are bouncing around again. Because each time the ball machine is turned on is a new chance for any ball to pop up, you can see the same number two, ten, or one hundred times in a row. The probability of it appearing in the next draw remains the same. The balls don't get more reluctant to play along the more often they make it out of the chute.

If for some reason you don't believe the laws of probability apply here—maybe because you think the government rigs the lottery, or whatever conspiracy theory catches your fancy—statistical analyses done by Torrey Pines High School student Albert Chen should resolve the question. In a paper he published with a professor at San Diego State University, Chen examined whether the low-frequency strategy gave players an advantage in the California Lottery. It did not. The lottery is not rigged.

Chen, who was also his school's homecoming king, cautions, "The lottery is simply a game of chance. No matter which strategy is used, the probability of winning a large cash prize is extremely low. We suggest that no one plays the lottery unless it is done to have fun or to support education." Well-rounded advice from a well-rounded young man.

The *Good Morning America* reporter provided her own cautions. She warned viewers to expect some losses, even if they used Lustig's strategies. But, she proclaimed, they should never get discouraged. Lustig certainly doesn't. She said, "And though he's won big, he loses, too. But he never stops playing." So, by implication, she encourages us to continue playing the lottery because someday we might win. This piece of advice falls into a third category: mathematically correct, but unhelpful. It is true that you can't win the lottery if you don't play. But it is not true that if you keep playing, you will eventually win.

The lottery is not a game that can be solved with math. All you need is luck. Lots of luck, given the odds. But math can, in turn, help you see that you will not likely win the lottery and might be better off saving your money or buying a candy bar with it. The problem with using math to win the lottery is that the lottery is totally random. In contrast to the lottery, games that mathematicians can solve do not operate entirely on chance. They involve players making logical moves in response to the structure of the game and the actions of other players.

Math doesn't ruin all games that involve logic. Sometimes mathematicians create solutions to games that humans can't use. Game theorists don't necessarily set out to create game solutions that are practically useless for human players. But with powerful computers at their behest, they often end up doing so.

JONATHAN SCHAEFFER WANTED TO solve checkers so that he could beat Marion Tinsley, the best checkers player who has ever lived. Between 1955, when Tinsley won his first checkers world championship, and 1994, Tinsley lost only nine matches. Tinsley could see the consequences of moves further ahead than any other player. In one game, Schaeffer calculated that Tinsley would have had to see a full sixty-four moves into the future to know that a move he made was a good one. Was this incredible ability good luck or something else? According to Tinsley, it was divinely inspired, a special gift to him from God. "God gave me a logical mind," he once said.

Schaeffer was obsessed with beating Tinsley. "Tinsley was the Mount Everest we wanted to scale," Schaeffer told the *New York Times* for Tinsley's obituary in 1995. Schaeffer could never get good enough at checkers himself to beat Tinsley. But he knew something Tinsley did not: how to program computers. So, Schaeffer built a checkers-playing computer and challenged Tinsley to a match.

The showdown was set for 1994. The first game ended inconclusively, in a draw. So they had a rematch. Again, the game ended in a draw. They played again. In the end, man and machine played six games. All six ended in draws. A seventh match was in the cards . . . but then Tinsley had to withdraw for the saddest and most human of reasons. He fell ill, and eventually passed away from pancreatic cancer.

Still obsessed with his desire to beat Tinsley, even though Tinsley was dead, this lack of finality in the battle between Tinsley and Schaeffer's computer drove Schaeffer to try to prove that his computer *could* have beat Tinsley, even though it never did. But how to beat a dead man at checkers? The only surefire way to prove that he and his computer could have beaten Tinsley was to turn to game theory. Schaeffer needed to solve checkers.

You may have heard of DeepBlue, the computer chess master, or Watson, the machine that plays *Jeopardy!* better than even Ken Jennings. These computers are impressive. But the machine that Schaeffer eventually built, which he named Chinook, surpassed both of these more famous computers. Chinook may only be able to play checkers, but it plays perfect checkers. DeepBlue and Watson only play really, really good chess and *Jeopardy!*, respectively. That's because the program Chinook uses follows the mathematical solution to the game. Like tic-tac-toe, checkers is solved. Chess and *Jeopardy!* aren't. Schaeffer's obsession is satisfied.

You might think of checkers as a lame game for kids, similar to tic-tac-toe. But unlike tic-tac-toe, checkers involves a significant amount of strategy. It is actually a complex game. You probably have an intuitive sense for what makes a game complex. Are there lots of pieces? How much information can players hide from each other? How likely is a seven-year-old to start making up silly rules for it because she doesn't understand the real rules and gets bored? On these measures, checkers isn't a frustratingly complex game. But neither is it easy.

How mathematicians think about a game's complexity is similar to the way a layperson might. But they measure its complexity in a way that no regular player of games ever would. Mathematicians calculate the exact number of possible outcomes in a game and then try to determine how difficult it is for a player to choose a sequence of moves that ends in victory.

Calculating the number of possible outcomes of checkers is an immense task. It's so immense that chances you've ever played two identical games of checkers are remote. What happens in a game depends on the whims of the players. Nonetheless, there are only a certain number of plots that can unfold. Checkers players may feel as though anything is possible, but their choices are bounded. Granted, the boundary encompasses so many options that for any human player it may as well not exist. But the exact number of outcomes of a game matters to mathematicians looking to solve it. Mathematicians need to know everything that can happen in order to make a victory plan for all possible situations.

To help manage this immense task, mathematicians build game trees, or maps of all the choices that players could make and their consequences. What players experience as racking their brains while imagining the outcomes of a move mathematicians see as a tree.

The trunk represents the beginning of the game. As players make choices, the tree grows branches. One branch represents the first player moving her leftmost piece two squares forward. An adjacent branch shows her moving her second leftmost piece instead. And so on. Each branch branches again as players make more choices and move more pieces. Different branches lead to different possible outcomes for players. Good moves tend toward victory; poor ones toward defeat. A game's tree holds all the information you'd ever want to know about a game. It contains all possible games that could ever be played. For most games, that's a big tree with a lot of branches.

So, according to mathematicians, how complex is checkers? Mathematicians call it a moderately complex game. With 500,995,484,682,338,672,639 different possible situations, you might expect them to call checkers a very complex game. Believe it or not, however, this is a moderate number of possible situations compared to some other games, such as chess or the popular Japanese game Go.

If you have a complete tree for a game, you have the game's solution. To win, all you have to do is make the choices along the branches that end in victory. After your opponent makes a move, you can use the tree like a map to see what happens if you respond in different ways.

But making a map of five hundred quintillion different choices is no small endeavor. Even Tinsley, with his supernatural foresight, could not possibly hold such a map in his mind, let alone construct one. Building a flawless map of checkers—solving the game—requires more mind power than any single human possesses. And more time than any human brain has on Earth.

It takes, in fact, about two hundred computer processors eighteen years. At least that's how long it took Schaeffer's army of computers. Schaeffer set his checkers-solving program named Chinook into motion in 1989, and it finally solved checkers in 2007. According to mathematicians, checkers is now as boring as tic-tac-toe. That is, if you are a computer.

And what did Schaeffer learn? Could Chinook have won a game against Tinsley?

The conclusion was anticlimactic. *Maybe* the computer could have won. That's because if two parties both play perfect checkers, the game will always end in a draw.

If Tinsley was programmed by God, as he claimed, he might still have found a divinely inspired way to have beaten Chinook. For Chinook was programmed by Schaeffer, a mere mortal. Surely, divinely played checkers is better than perfectly played checkers.

When Schaeffer solved checkers in 2007, mathematicians around the world took notice. It was by far the most complex game ever solved. Schaeffer's solution to checkers was a milestone.

But aside from the impressive amount of work involved, what was the significance of Schaeffer's solution? What did his solution mean for checkers players? For humankind in general?

In some ways, his solution meant little to checkers players. No human player—except, possibly, Tinsley and his superhuman brain—could follow Chinook's program. It's much too complicated. It doesn't come with a strategy book. So, unless you find yourself playing checkers against a robot programmed with Schaeffer's solution, in which case you might as well throw in the towel, don't expect your checkers game to be affected by Schaeffer's accomplishment. Human checkers goes on as normal, even though the solution to checkers exists in the depths of Chinook's processors.

But mathematicians don't solve games just to build computers that can maybe beat world champions. They solve games to further game theory, that branch of mathematics that attempts to make sense of how people behave during conflict. To a game theorist, checkers and tic-tac-toe are games. But so are World War I, nuclear proliferation, and global trade. Mathematically speaking, a game is simply a situation in which at least two parties compete for the best outcome in a shared situation. Competition takes the form of making choices, or game moves. Whether the parties get a payoff or have to pay up depends on whether their choices result in winning, losing, or something in between.

Checkers is obviously a game for at least three reasons. First, two people play it, but only one can win. Second, on any turn, players choose from a set of moves, or places to put their red or black chips. And, finally, depending on what moves they make, they may win or lose the game.

World War I was also a game—but much less fun for the participants than checkers. Especially for the frontline soldiers, who were the equivalent of checkers pieces. Countries competed to conquer territory and take valuable resources. At any time, a country could make a "move": advance an army into another country, build new weapons, or gallantly call a truce in the name of cooling things down. Depending on what each country did, the world plunged deeper into conflict or moved closer to peace. Some players won, some lost. Most both won and lost. A payoff of fearsome strength for any nation also cost that nation the deaths of many soldiers and civilians.

Mathematicians who solve games get to challenge world champions to high-profile checkers tournaments. But they also get to develop the theories on which nations base their wartime strategies. Mathematicians solve games like checkers to gain insights into these more important games. One of the primary goals of game theory is to uncover a mathematical structure within the chaos of conflict so that those responsible for making life-or-death choices can make those choices wisely.

Does Schaeffer's solution to checkers help with this important task? In some ways it does. Chinook was a computing breakthrough. Problem solvers from all fields who rely on powerful computers can benefit from Schaeffer's work.

But this is about as useful as Chinook gets for solving world conflict. That's because checkers and World War I differ in a fundamental way. It's not that checkers is fun and World War I wasn't. The fundamental difference between games like checkers and real-world games like war is that all it takes to solve checkers is a powerful computer. War, however, has added complexity that no computer can handle.

In many ways, war is more like the lottery than it is like checkers. While logic directs some aspects of world leaders' behavior, much of what happens in war is governed by chance. To capture this aspect of chance, mathematicians group games into two main categories: games they call "perfect information" and those they call "imperfect information."

Checkers and tic-tac-toe are perfect-information games. In these games, all of the information a player needs to make a perfect decision is on the table, even if in the case of checkers you need a computer to analyze it. Nothing is left to chance. Players can't hide information or make decisions in secret. If one player seems sneaky, it's because the other player isn't observant enough to follow what's going on. In checkers, making the perfect decision basically comes down to running through all five hundred quintillion different possible game situations to choose the perfect branch of the game tree. All of the information is available to use—if you have the capacity to process it.

Checkers is so complex that, for our puny human brains, it may as well have hidden information. But games with perfect information are much easier for mathematicians to solve than games with imperfect information. Powerful computers can navigate the immense checkers game tree. So, solving a perfect-information game comes down to building a sufficiently powerful computer. This is not a simple feat, but it is relatively straightforward.

But the straightforwardness of solving perfect-information games comes at a cost. Perfect-information games are unrealistic. The possible outcomes of most real human conflicts cannot be fully mapped. That's because most real conflicts involve information that players can't access.

In most real conflicts, information essential to making choices can be kept secret by players or is a matter of chance. Conflicts like these are called imperfect-information games. In these games, having a perfect memory isn't enough to play the game perfectly. Poker is an excellent example of an imperfect-information game. So is war.

In poker, the imperfect information can come in two forms. First, players are dealt cards at random. Nothing, not even a computer, can determine for certain which cards a player will receive.

Second, players can hide some of their cards. In Texas hold 'em poker, for instance, players initially receive two cards, face-down. Then the dealer places five cards face-up in a community pile. Everyone can see the community pile cards, which provide some useful information for players trying to decide whether their hand is strong enough to stay in the game. But no one has all the information they need because of the two hidden cards. Players can never compare their hands to the hands of their opponents. So, even the most well-considered decisions are guesses.

When information is imperfect, players can manipulate each other. They can bluff. They can posture. The complexity introduced by secrecy and chance is thus compounded by emotional exploitation. So, while the game tree for checkers has well-defined branches, the game tree for poker is cloudy. What players see is something like a tree viewed at a distance by a near-sighted person not wearing glasses: a fuzzy collection of clumps of branches grouped together by their most visible features. States of the game with the same public cards but different hidden cards look the same to a player. The player cannot know with certainty the actual state of the game because there are many different possibilities depending on what are the hidden cards of the other players. All of those possible game states clump together on the tree and blend into an indistinguishable mass.

So, even though checkers has a game tree with a thousand times more branches than some forms of poker, only the simplest forms of poker have been solved. In 2015, some of Schaeffer's colleagues solved heads-up limit Texas hold 'em, a version of poker with only two players.

But if you found Schaeffer's solution to checkers unsatisfying because perfectly played games always end in a draw, you'll be even less satisfied with the solution to heads-up limit Texas hold 'em. Cepheus, the computer that plays perfect poker, doesn't just tie. It often loses. In the short run, at least.

If you play a limited number of poker games against Cepheus, there's a good chance you'll beat it. Professional poker player Michael Shinzaki found this out when he played two hundred hands against Cepheus and walked away ahead. In fact, the probability of being ahead of Cepheus after one hundred hands is nearly fifty percent.

It makes sense that Cepheus would lose sometimes. Even computers cannot see through the backs of cards, read minds, predict the random dealing of the cards, or tell the future. Eventually, however, Cepheus starts to come out ahead. Cepheus's solution to poker relies on the fact that, in poker, winning is cumulative. It matters less if you win individual games than if you amass winnings in the long run. For most of us, "long run" means ten or twenty hands at a time. We get tired. Cepheus doesn't.

If you manage to play thirty thousand hands against Cepheus, your probability of coming out ahead drops to five percent. It continues to fall from there. After millions, billions, and quintillions of hands, more than a human could physically play in a lifetime, the probability of beating Cepheus is so small as to be virtually nonexistent. And this is what mathematicians mean when they say that they've solved heads-up limit Texas hold 'em. Their computer loses individual games. But it wins the ultimate poker game of attrition.

Cepheus does this by making fewer mistakes than its human opponents. Cepheus uses its inhumanly immense memory to run a program called the Regret-Minimizing Algorithm. This algorithm does exactly what its name suggests. Whenever Cepheus makes a move it later regrets, it learns to not make that move again. Future regret is minimized—but not eliminated. For, despite its impressive memory, Cepheus will never stop making mistakes. Unlike Cepheus, however, human players can make the same mistakes twice and often do. So, Cepheus eventually comes out ahead.

The solution to heads-up limit Texas hold 'em is less satisfying than the solution to checkers. Cepheus's way of slowly edging ahead is not what we usually think of as a *solution*. We think of solutions as always working. Not almost there or always a little off. In the battle of human reason versus chaos, the ultimate weapon that humans currently possess, a logic-wielding computer, does not provide decisive victory. A chance of losing is still a chance of losing, no matter how small.

For better and for worse, Cepheus is closer to the kind of solution we should expect for real-world games, games like war. Imperfect-information games like poker more accurately reflect the real-life "games" that humans play with one another for dominance and survival than do checkers and tic-tac-toe. In 1919, when the Allies and the Central Powers negotiated the Treaty of Versailles to end World War I, they didn't put all their cards, red and black pieces, or X's and O's on the table. They kept some information to themselves. And some of what happened afterward was left up to chance and to the connivance of the parties. Who could have predicted, for instance, that an outbreak of the flu in 1919 would have devastated soldiers and civilians on both sides of the war? Who could have anticipated that shortsighted "beggar thy neighbor" economic policies would be adopted by the most powerful countries, and that these policies would eventually lead to the Great Depression of the 1930s?

As a result, we cannot expect that mathematicians will solve games like war as completely as they have solved tic-tac-toe and checkers. Their solutions will always be imperfect. As imperfect as the information and the people they have to work with.

The Evolution of Trust

Mathematicians' solutions to real-world games may be imperfect. But they shape global policy. Policy makers are always looking for objective (or, at least, objective-looking) rules to help them shape successful policies. Mathematicians' predictions about how people in conflict behave can mean life or death for soldiers. So, how do mathematicians deal with the imperfect information inherent in human conflict?

For a window into this question, let's look at how mathematicians solve one of the simplest human conflict problems: the Prisoners' Dilemma. For game theorists, the Prisoners' Dilemma is to war as tic-tac-toe is to checkers or poker. It's a game at its most basic. But it provides insight into how game theorists attempt to solve much more complex human conflicts. In fact, game theorists use the Prisoners' Dilemma and its solution to model human behavior during war.

Here's how it goes. You and a friend are arrested on the suspicion that the two of you robbed a bank. It doesn't matter if you actually robbed the bank. It also doesn't matter if your friend did, or if you know either way. What matters is that the police have sequestered you both in different interrogation rooms. The police and the prosecutor want to close the case. So, they offer you and your friend the same deal.

If you confess that the two of you committed the robbery and your friend stays silent, you go free. The prosecutor will use your testimony to ensure that your friend spends ten years in jail. The same goes for your friend. If your friend confesses and you stay silent, you have to spend ten years in jail while your friend goes free.

But here's the catch: if neither you nor your friend confesses, you will both go to jail. The prosecutor has enough evidence to send both of you to jail for one year on a lesser charge. Also, if you both confess, you will both go to jail for the robbery. But because two confessions make the prosecutor's job easier, you'll each serve only five years.

Because you're in different rooms, there's no way for either of you to know what the other will do. What should you do? Table 1 outlines the possibilities.

While you think, I'll establish that this is actually a game. It meets what we have described as the three criteria for a game. There are two players. Your choices determine what payoff you get. In this game, "winning" is defined as not going to jail. "Losing" is defined as getting the maximum penalty, ten years in jail. Going to jail for less time, either one or five years, falls somewhere between winning and losing.

This isn't a perfect-information game, because you don't know what your friend will do. But the game is simple enough that you can think through all possible scenarios for you and your friend. Just like for tic-tac-toe, we don't need an enormous computer to solve this game.

Let's see if we can solve it. The only way for you to "win," or not go to jail, is to confess while your friend stays silent. Furthermore,

Table 1. Prisoners' Dilemma

	You stay silent	You confess
Friend stays silent	You: *1 year* Friend: *1 year*	You: *Free!* Friend: *10 years*
Friend confesses	You: *10 years* Friend: *Free!*	You: *5 years* Friend: *5 years*

The possible jail sentences for you and your friend depend on whether you confess or stay silent. While you both minimize your jail time if you both stay silent, staying silent risks spending the maximum time in jail—ten years.

the only way for you to not "lose," or go to jail for ten years, is to not stay silent while your friend confesses. You might lose if you stay silent. And if you stay silent, you definitely cannot win. So, the solution is that you should confess.

But this doesn't mean that you are certain to win the game. That's because your friend will also make the move most likely to result in winning, namely, to confess. Your friend wouldn't do something that could lead to total loss. As such, you can't rely on your friend to stay silent.

According to the best possible solution to this twisted game, you both will confess. The outcome isn't great. You probably both go to jail for five years. But at least you didn't go to jail for ten years! If you play the game according to the best possible solution, you probably won't win outright, but at least you won't lose. Just like the solutions to tic-tac-toe and checkers, the solution to the Prisoners' Dilemma eliminates your chance of losing outright while preserving your chance of winning. It tells you what you should do no matter what your friend does.

This kind of solution to a game is called a Nash equilibrium. A Nash equilibrium is a solution to a game that provides the most logical choice to each player in a situation in which neither knows what the other is going to do. The choice provided is the most logical irrespective of what the other player actually does. In a Nash equilibrium, neither player has reason to change their choice, even if they find out the other player did not make the most logical choice.

Nash equilibrium is so important to how game theorists think about real human conflicts that the man who invented it, John Forbes Nash Jr., won a Nobel Prize for developing it. Nash equilibrium is powerful because it explains, simply and logically, how people should act in real conflicts. Without Nash equilibrium, the Prisoners' Dilemma is full of terrifying unknowns. But with it, you have control. It doesn't matter that you don't know what your friend will do. You can have confidence in your decision. Even without perfect information, Nash equilibrium guarantees that you won't lose. It also gives you a solid chance of winning.

The Prisoners' Dilemma is remarkably similar to the situation that opposing nations find themselves in during war. The leaders of a nation must decide: Should we fight for victory at all costs, or should we settle and negotiate peace? Like you and your friend alone with the police, each side in a war must make decisions without knowing what the other will do. But both sides can think through the possible scenarios and make their best plausible choices.

"Winning" in the game of war is total victory for one side and total defeat for the other. Total victory is usually impossible, however, if both sides fight to the end. War takes its toll, even if your side comes out victorious. Total war for total victory can lead to a situation in which your win is not worth what you have lost in the fighting. But like the Prisoners' Dilemma, war often gives both sides an option that minimizes their losses: suing for peace. Entering into a peace settlement involves making compromises, however. It also involves some risk. What if the other side breaks faith during the negotiations and strikes when the leaders least expect it? Or what if the other side uses the peace merely to prepare for more war? The faithful nation could lose everything. Table 2 shows the possibilities for war and resembles the table we drew for the Prisoners' Dilemma.

Table 2. Possible Outcomes of War

	You sue for peace	**You fight to the end**
Enemy sues for peace	You: *Make some concessions* Enemy: *Makes some concessions*	You: *Victory!* Enemy: *Total loss*
Enemy fights to the end	You: *Total loss* Enemy: *Victory!*	You: *Suffer war, and then make some concessions* Enemy: *Suffers war, and then makes some concessions*

Just as in the Prisoners' Dilemma, warring nations concede at their own risk. The most stable outcome is for both sides to fight until they must negotiate.

And the Nash equilibrium solution is the same. Because both sides risk losing everything and give up any chance of total victory if they sue for peace, both sides will fight until they cannot stand to fight anymore. Then one side may surrender and probably lose whatever they were fighting for. The other side will proclaim victory but will likely have lost much in the fighting. And, sadly, this is what often happens during war. Opposing nations kill endless streams of soldiers, destroy each other's economies, and bomb civilians, until the war of attrition ends. All because the risks of suing for peace seem too great.

Game solutions based on Nash equilibrium may be reliable. But they are also depressing.

If you base your decision-making on the Nash equilibrium, events such as the Christmas Truce of 1914 aren't just surprising. They are mathematically incorrect. In spontaneously suing for peace before the war took its full course, the British and German armies did not follow the solution to the game that is mathematically prescribed by the Nash equilibrium. They did something that math has generally said you should never do when playing a game: they *trusted* each other.

But everything turned out fine. Both sides stopped fighting, at least for long enough to celebrate Christmas. Does that mean that both sides were illogical and naive? Nash would say yes. Trusting your opponent is always illogical. Mathematicians Robert Axelrod and Nicky Case would say no. Sometimes, these mathematicians argue, trusting your opponent is the most logical thing to do. Sometimes, peace is the solution to games of war. Perhaps there is a way—a logical and not merely sentimental way—to give peace a chance.

To understand how this works, let's return to the original Prisoners' Dilemma problem. Here, as in the game of war, we assumed that trust is illogical. But the Prisoners' Dilemma game differs from the war game in that the prisoners are friends, not enemies. The Nash equilibrium solution to the Prisoners' Dilemma essentially makes you and your friend into enemies. You must be prepared to put your friend in jail for ten years while you walk free. And you must assume that your friend will do the same to you.

But does this make sense? Remember, you are *friends*. You know each other well. You ostensibly like each other. And, presumably, you have a history of trusting each other and having that trust validated. Otherwise, how could you be friends? Yes, in this situation, if you trust your friend and your friend turns out to be not such a great friend after all, you go to jail for ten years. You spend the next ten years agonizing over how gullible you were, and how you should have known that it's a dog-eat-dog world out there. The consequences of misplaced trust are dire. But reflect on your years of friendship. Would you actually throw your friend under the bus for your own possible benefit? Could you live with yourself if you did? Would your friend do the same to you?

If you and a genuine friend were really in this situation, I bet neither of you would confess. And, in fact, math developed since Nash's time shows that this is the right decision.

Game theorist Robert Axelrod first explored the idea that trust might be logical in competitive situations while collaborating with an evolutionary biologist named William Hamilton. Biologists had long assumed that evolution was a lot like Nash's version of war: a survival-of-the-fittest battle to the top, in which cooperation is a sign of weakness. But Hamilton and Axelrod began to notice situations similar to the Christmas Truce cropping up in the natural world. Sometimes, it seemed, species that should compete cooperated. And trust made them evolutionarily stronger.

For example, take lichen, a scaly growth often found on trees. A lichen is actually two species in one, a fungus and an alga living together. Logically, the fungus and alga should compete for resources. They live on the same small tree branch. Light, water, and nutrients are scarce. But in a lichen, neither species can survive without the other. They evolved this way. Trust and cooperation make them stronger.

Axelrod hypothesized that the fungus and alga were able to cooperate, despite all signs pointing to their being competitors, because they had a long history together. Like you and your friend, the fungus and alga have proved themselves trustworthy over the years of their relationship. Axelrod theorized that when two parties in a game know each other and are likely to play the game again, it makes more sense for them to trust each other than to throw each other under the bus. They've had several interactions in which both acted for mutual benefit. So, in each subsequent interaction, neither fungus nor alga assumes that the other will choose solo victory over collaboration. Doing so would be illogical. Trust makes the most sense.

How does Axelrod's theory change the outcome of the Prisoners' Dilemma? Let's play a game and find out.

Mathematician Nicky Case turned the Prisoners' Dilemma into an actual game. Case calls the game "The Evolution of Trust." In it, Case explores what happens when you make different assumptions about trust. If you have a computer, you can play it yourself. But I'll work through a few scenarios here.

The Evolution of Trust is a two-person gambling game. Each person gets one coin per round. If you put a coin into the machine, your opponent wins three coins—and you lose your coin. If your opponent puts in a coin, the opposite happens. You get three coins, but your opponent loses her coin. In turn, if you both put in your coins, you each net two coins. In each round, you can choose to do what Case calls "cooperate," put in your coin. Or you can "cheat," not put in your coin. Table 3 shows the possibilities.

As in the Prisoner's Dilemma, this game initially encourages you to not trust your opponent. No matter what your opponent does, you shouldn't put in your coin. You can't lose if you keep yours. And if your opponent is naive enough to put in her coin, you might win big.

But in Case's game, you play many rounds. Unlike the traditional Prisoners' Dilemma scenario, the goal isn't to minimize your losses and maximize your chance of winning big in one game. In Case's game, you want to find a strategy that you should always follow over many games with the same player. Does changing the number of games you play change the solution to the game?

Case lets you explore this question. In Case's game, you play multiple rounds against different opponents. Each opponent acts in different ways. It's hard to know your opponent's strategy on the

Table 3. Possible Outcomes of the Evolution of Trust Game

	You put in a coin (cooperate)	**You keep your coin (cheat)**
Opponent puts in a coin (cooperates)	You: *Win two coins* Opponent: *Wins two coins*	You: *Win three coins!* Opponent: *Loses her coin*
Opponent keeps her coin (cheats)	You: *Lose your coin* Opponent: *Wins three coins!*	You: *Nothing* Opponent: *Nothing*

Whether and how much you win depends on whether you and your opponent cheat or cooperate. In a single round of the game, the greatest payoff and least risk of losing come from not trusting your opponent or acting in a trustworthy way yourself.

first round. But if you play enough rounds, you can learn what your opponent does. Then you can adjust your strategy to get the best possible outcome.

One opponent plays as Nash would—never puts in a coin, or always cheats. A second opponent naively always cooperates and puts in a coin. Against both of these players you should do as Nash suggests and always cheat. If you cheat against a player called Always Cheat, you both lose nothing. If you cheat against a player called Always Cooperate, you win big. But the situation in most games is not stable because other opponents adjust what they do on subsequent rounds based on your behavior. As Axelrod puts it, their strategy evolves as they learn about your relationship. What you should do in response to their changes in strategies depends on the type of player you are facing.

For instance, one player called Grudger generally cooperates. But as soon as you cheat, Grudger never trusts you again. From then on, no matter what you do, Grudger always cheats. Grudger does not want to be taken advantage of. This is a situation in which cooperating would be your best strategy because then you would both earn coins, albeit slowly. If you get greedy and cheat to win big one time, you won't win again. The best you will be able to do thereafter is continue to cheat and thereby not lose.

Another player is called Copycat. Copycat also starts by cooperating. But Copycat isn't quite so quick to mistrust. Whatever you did on one round, Copycat will do on the next. So, if you cheat on the first round, Copycat will cheat on the second round. But if you cooperate on the second round, Copycat will cooperate on the third round. And so on. Copycat likes to give you a taste of your own medicine. But Copycat will also trust you if you are trustworthy. Again, as with the Grudger, cooperating would be your best strategy. You will both slowly earn coins.

So, against both Grudger and Copycat, Nash's solution, always cheating, gives you one round of victory. But continuing to cheat causes each player to cheat. The best you can do if you follow Nash's solution is win two coins on the first round, and nothing from then on. Against Grudger and Copycat, it is in your best interest to ignore Nash and cooperate. Cooperating gives both you and your opponent two coins on each round. If you maintain this trusting relationship, you can win as much money as you want. In these situations, cooperating is clearly better than cheating. You and your opponent can be like a fungus and an alga, living in lichen-like harmony forever.

There is no way to know at the beginning what your opponent will do. After all, this is an imperfect-information game. If by chance your opponent is Always Cheat and you cooperate on the first round, you'll lose a coin. Even if you continue to cheat from then on, you'll still have lost something. This raises the question: What should you do if you don't know who you are facing, an Always Cheat, an Always Cooperate, a Grudger, or a Copycat?

To help answer this question, Case holds a tournament that includes all of the players we have described. Always Cheat, Always

Cooperate, Grudger, Copycat, and another player named Detective play ten rounds against each other. Detective's strategy is to play the first four games while assessing what his opponent does. Detective's play in those first four games is as follows: cooperate, cheat, cooperate, cooperate. If his opponent cheats, Detective acts like a Copycat for the rest of the game. If his opponent doesn't cheat, Detective takes advantage of his opponent and always cheats for the rest of the game.

Case adds up their winnings after ten rounds. Who do you think does the best?

If you think like Nash, you might guess that one of the less trusting, more self-serving characters comes out ahead. Grudger's strategy is cautious. Maybe that's to her advantage. Detective's strategy mixes Copycat's learning with Always Cheat's thirst for blood. Maybe he's being smart. Always Cheat's strategy was proven best in the Prisoners' Dilemma. Maybe we shouldn't mess with a good thing. Always Cooperate is naive. Copycat, too. Certainly, a player like Always Cheat or the Detective must be able to take advantage of her trusting nature.

But one player significantly outperforms them all. By the end of the tournament, Copycat has the lead, with fifty-seven coins. The other players are at least ten coins behind.

The moral of this story is: when you have the time to build a relationship with your opponent, trust wins. Even when the situation seems competitive, you don't win by competing against the other player. You win by joining the other player in competing against the game. If you both cooperate, you both win coins. Whoever is running the game has to repeatedly pay the two of you. The most logical approach for both players, therefore, is to be trustworthy. Math says that so long as your opponent cooperates, you should, too. And trust can evolve.

I like this solution better than Nash equilibrium. And not just because it makes me feel optimistic. I think it's more realistic.

People who find themselves in conflict in the real world rarely get there without some prior relationship. That is true not only for individuals but for groups of people and even countries. Sure, you never know what your opponent will do. But real-world conflicts are rarely true imperfect-information games. It is most often the case that you will have a history with the person. You can generally know at least some of what your opponent has done in the past. That information is valuable. If your opponent has proven untrustworthy in the past, then you should consider cheating. Don't be naive. But if your opponent has a history of being trustworthy, why not take a chance on trusting? And if you and your opponent have a history of trust, why not act upon it?

AXELROD AND CASE'S realization that sometimes it's logical to trust someone who is ostensibly an opponent has implications beyond making us feel better about the world. Remember, mathematicians use game theory to model how people behave in real-world dire situations. They use those models to make important decisions—decisions that could mean life or death for people involved. Having the wrong models could mean making the wrong decisions.

Nowhere is it more important to have a good model than during an unpredicted, quickly evolving emergency, such as an epidemic. Epidemiologists rely on mathematical models of human behavior during epidemics to make time-sensitive decisions, such as where to send doctors and valuable supplies. But how will people act during epidemics? Will they behave as Nash might predict, caring for themselves over others? Or will trust and relationships come into play? Will they see each other as competitors for scarce resources or as colleagues with mutual interests working toward the common good?

Mathematicians' models for behavior during epidemics were put to the test in one of the largest epidemics ever: the Corrupted Blood Epidemic of 2005. Around four million were infected. It threw a complex society into chaos as officials scrambled to contain it. Eventually they succeeded, but not before many died and game theorists' ideas about human behavior were irreversibly challenged.

Never heard of the Corrupted Blood Epidemic? That's probably because you don't play World of Warcraft, a computer game involving a host of imaginary creatures. Players from all over the world play against each other as these creatures. Corrupted Blood wasn't a human epidemic. It was an epidemic accidentally released into the cyberworld by a software update gone awry. The epidemic was pure happenstance, at least as far as the humans involved with the game were concerned. As such, it provided the safest possible setting to test mathematicians' and epidemiologists' understandings of the ways people behave in a crisis.

On September 13, 2005, approximately four million World of Warcraft players were exposed to the full-blown epidemic. Corrupted Blood infected characters of all species. Orcs, dwarves, night elves, and pandaren, pandas with human hair, all succumbed. The players, glued to computer screens around the world, did not understand what was happening. Nothing like this had ever occurred in World of Warcraft. Some characters recovered, some died, and many others spread the disease to distant corners of the World of Warcraft universe. Even the game's developers hadn't anticipated the reach of the disease. Stopping the epidemic would require more than a few keystrokes.

Those unfamiliar with World of Warcraft might think that the Corrupted Blood Epidemic could be stopped if everyone just took a break from the game while the game's developers found a fix. But most fans are die-hard enthusiasts. They couldn't take that break. They had to heal their characters before it was too late. For them, a World of Warcraft epidemic was nearly as serious as an epidemic in the non-virtual world.

What would you do if an epidemic like Corrupted Blood were ravaging the real world? Lock your door and bar the windows? Hoard supplies, leave friends and family to sink or swim, and protect your immediate household from the infected masses? Or risk your life trying to help those affected? What if one of the people needing help was a close relative across town?

These are precisely the questions that mathematicians Nina Fefferman and Eric Lofgren sought to answer by studying players' reactions to the Corrupted Blood Epidemic. They published their findings in a respected medical journal, *The Lancet*. Medical journals don't typically address virtual medical problems contracted through software updates. They certainly do not deal with illnesses endemic to virtual orcs and elves. But Fefferman and Lofgren's findings were relevant to the journal's main concern, human health. What bridged the two seemingly distant worlds of medicine and online games was math.

To mathematicians, how people behave during an epidemic is a game. An epidemic is on its face a competitive situation because the two groups of people involved—the sick and the healthy—can be seen as having opposite needs. They have the same basic goals, for the epidemic to end and to stay alive until it does. But their ways of achieving their goals could conflict with each other.

During an epidemic, it's in the best interest of healthy people to stay far away from the sick. If healthy people stay away from the sick and don't catch the disease, the epidemic is more likely to dissipate than if the healthy people continue to interact with the sick people. But it's in the best interest of sick people to find a healthy person to take care of them. If sick people don't get care, it is more likely they will die. Therefore "winning" for one type of person could be "losing" for the other.

Many mathematical models that predict what people will do during epidemics presume that people will act selfishly. The Nash equilibrium's solution to the Prisoners' Dilemma did the same. If the healthy and sick are competing to avoid dying from the disease, wouldn't they each take the action that is best for themselves, regardless of what happens to those around them? The assumption is that people won't care about the people they're competing against—each person will act for themselves.

But in some situations that have competitive components, such as trying to stay alive during an epidemic, it turns out that how people behave is much more complicated. The problem is that classic game theory models don't explain why healthy people sometimes risk their lives to care for the sick. It also doesn't explain why some sick people separate themselves from others in heroic efforts to reduce the spread of disease. As it turns out, real people don't always try to win. In real-life epidemics, as in the Prisoners' Dilemma, relationships matter when people decide what to do.

And it's important that mathematicians and epidemiologists know in advance what people will do. That's because epidemics are spread by human contact. Public health officials need to be proactive to stop epidemics, so they need to know who will reach out to whom. If people behave as Nash predicts, with the healthy isolating themselves and the sick seeking care, the disease will take one trajectory. But if people try to help or protect those they care about, possibly including strangers, the disease may spread in a completely different way.

So, how do people act during epidemics? What factors affect whether people will act altruistically or selfishly? Perhaps it's how awful the disease is. Maybe it's how close people's relationships are in the community in which the disease is spreading, or how medical expertise is distributed. These are important questions for mathematicians to answer if they're going to build accurate computer models about epidemics. But it's hard to study real people in these circumstances. So, Fefferman turned to World of Warcraft and its Corrupted Blood Epidemic for answers.

Fefferman, a professor in the biology and math departments at the University of Tennessee, uses math to model how people behave during epidemics. She typically collects data while epidemics are happening, examining the relationships between human behavior and the spread of disease using computer models. Sometimes Fefferman also uses models that other people developed to test claims about what might happen during a real epidemic. But her most groundbreaking research involves inventing models herself.

World of Warcraft was a good site for Fefferman's behavioral research because the characters in the game could get sick—and die—without hurting any actual humans. But the characters' behavior was completely controlled by players who cared about their characters' health.

World of Warcraft was, of course, just a game. What happened in the game had no direct bearing on real life. Fortunately for the players, the company that makes the game was eventually able to stop the disease and reset much of the damage. But for many of the players, the epidemic was a real-life experience for them.

World of Warcraft players can be extremely serious about the game. They spend hours playing each week, taking their characters through elaborate quests that lead to personal glory and advance the plot of the game. In doing so, players develop relationships with each other. A sense of community develops in the game. During the Corrupted Blood Epidemic, the altruism that grew out of community attachment seems to have greatly changed the course of the disease.

Sick characters, under the control of their human players, traveled around making others sick—sometimes because they didn't know about the risks, and sometimes because they were looking for help. Fefferman observed characters with special healing powers flocking to areas with lots of sick characters, trying to help. Some of the characters in World of Warcraft were less susceptible to Corrupted Blood than others, much as some humans are less likely to suffer from epidemics than others. But, like humans, characters who were less susceptible to dying from the disease but were carriers of it kept accidentally spreading the disease to weaker characters—especially when they rushed in to save those in need.

As a result, altruism didn't always help. Actions that may have seemed logical to a player trying to help his friend had implications beyond what anyone anticipated. In turn, altruism affected the spread of the disease in ways that led Fefferman to conclude that traditional game theory models weren't accurate. Sometimes altruism made things better, sometimes worse. In any case, game theorists needed to factor altruism into their models. Fefferman made new models that captured what she had learned from the Corrupted Blood Epidemic. Those new models are now in use around the world, helping us prepare for real epidemics when they strike.

But what Fefferman's work shows more than anything else is that in *real* imperfect-information games, ones that are far more complex than the Prisoners' Dilemma, mathematicians still have a lot to learn. When human behavior takes mathematicians by surprise, maybe the humans aren't the ones acting illogically. Maybe it's the mathematicians who don't understand.

A Mathematically Possible Peace?

In the depths of the winter of 1914, soldiers traded their guns for beer steins and raised them in toast to their enemies. The Christmas Truce did not last long. Eventually, the soldiers retreated to their trenches. They picked up their weapons and resumed the war. But was the Christmas Truce really as improbable as it seems?

Consider the Christmas Truce from the perspectives of Axelrod, Case, and Fefferman. It's hard to see at first, but the truth is that these soldiers had relationships, even across enemy lines. First, almost all of them were working-class conscripts whose lives had been similar before the war. None of them had chosen to be cannon fodder in a war started by the world leaders who were safe at home. Second, they had been playing the game of war against one another for six months. They were familiar with each other's actions and reactions. They knew that if you fire at me, I'll fire at you. But they also knew that if one side didn't fire, the other side might keep the peace. Mistrust had evolved between them. But so had trust.

As Case says at the end of their discussion of their game, the Evolution of Trust, "Game theory reminds us [that] we *are* each others' environment. In the short run, the game defines the players. But in the long run, it's us players who define the game." Game theory doesn't just tell us what to do in competitive situations. Sometimes it shows us how to transform what could be competitive situations into cooperative situations. Sometimes it gives us a choice. And in the brutal game of World War I, in the winter of 1914, the soldiers had a choice.

The soldiers could play the game of war as an Always Cheat and shoot at soldiers from the other side when they offered a temporary true. Or they could play like a Copycat and accept a Christmas Truce when it was presented. They had a choice to continue violence or copy peace. At Christmas, they chose to copy peace. With Case's model of a Copycat in mind, suddenly the Christmas Truce doesn't seem so mathematically improbable after all.

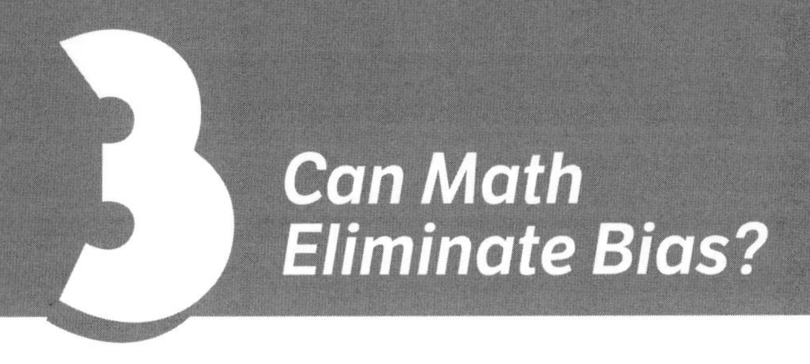

3 Can Math Eliminate Bias?

Math and the Problem of FAIRNESS

Justice by Algorithm

A man is escorted into a New Jersey courtroom. He's wearing the gray jumpsuit of the local detention center where he is being held on charges of drug possession. He has prior convictions. He also has a history of not showing up in court again after being released on bail. But now he has four children. Without his job, he cannot support them. And he can't keep his job if he's in jail.

If this scene had happened before 2016, the judge would have had little leeway in what bail he could set. The defendant would have left the courtroom facing thousands of dollars in bail, regardless of his ability to pay. And if he couldn't post bail, he would sit in jail for months, maybe longer, waiting for his trial. Judging by the tears in the man's eyes as he fearfully awaits his fate at the hands of Judge Ernest M. Caposela, who will set his bail, it seems that the defendant probably cannot afford much bail. If he goes to jail, he will lose his job. His family will be destitute. He and his family will have suffered severe harm even if he is ultimately found not guilty. And all of this suffering would be the result of his awaiting trial for a crime he may not have committed.

You might think this sounds like putting the punishment before the conviction. Punishment of the poor based merely on accusation. This does not seem fair. In several states around the country, including New Jersey, lawmakers have agreed. In these states, this man's life could turn out much differently. He has a better chance of keeping his job and caring for his family at home while waiting for his trial. And that's because of math.

THE MAN AWAITING HIS FATE at the hands of Judge Caposela is from New Jersey, a state that recently pioneered a new bail-setting method for people accused of nonviolent offenses. In New Jersey, since 2016, such defendants walk free without bail before their trials. They must check in regularly. But if defendants are judged not to be flight risks or threats to public safety, they can live in their homes, keep their jobs, and take care of their children before facing trial.

The new bail system depends on Judge Caposela's ability to reliably assess whether a defendant has a high probability of fleeing or committing a crime between release and trial. In criminal-justice speak, the latter worry is called recidivism. Recidivism is the chance a defendant or someone convicted of a crime will commit a crime in the future.

Many police officers and citizens in New Jersey worried that if they released people, even those charged with nonviolent offenses, without bail before their trials, crime rates would increase. They reasoned that where there is smoke, there is often fire. At least some of the people probably did commit the crimes for which they were charged. Releasing them without bail might be risky. How, they asked, can you determine risk unless a trial has already taken place?

Proponents of the new bail law counterargued that there are benefits in removing bail for most defendants in nonviolent criminal cases, and the benefits of the new system outweigh the potential detriments. Whether a defendant can pay bail is determined by how wealthy the defendant is. Rich people get out on bail. Poor people stay in jail. But there is no correlation between wealth and innocence or poverty and guilt. "There's nothing that says you can't be a serial killer and a millionaire," said Joseph E. Krakora, a New Jersey public defender. On the flip side, there's nothing that says you can't be innocent of the charges against you and living paycheck to paycheck. To some, the old bail system unfairly let the wealthy walk free while locking up the poor, regardless of their risk of recidivism.

New Jersey lawmakers believed their old bail system, in which poor defendants could not afford to post bail but rich defendants could, violated the concept that all defendants are innocent until proven guilty. The logic is that poor defendants awaiting trial in jail because they cannot afford bail might be getting punished for a crime they didn't commit. The old system also disproportionately affected defendants from minority groups, especially black defendants. New Jersey state officials, defendant advocates, and judges like Judge Caposela have all hailed the new system as fairer. In particular, they believe it evens the playing field for minority defendants who are less likely to be able to come up with bail.

But the worry about recidivism remains. Ability to pay bail might not be a *fair* way to determine whether it is risky to let a defendant await trial at home, but that doesn't mean it is better to eliminate bail altogether. New Jersey needed a procedure for assessing the risk of recidivism. This procedure needed to nail down a defendant's risk with calculator-like precision. In order to protect the public and gain the public's confidence in the system, the procedure had to produce reliable results each time, with computer-like regularity.

New Jersey turned to math. How better to produce reliable results than by calculating the probabilities of recidivism with math? New Jersey found its procedure in the algorithms of the Laura and John Arnold Foundation's Public Safety Assessment.

The Public Safety Assessment, or PSA, as it's known, uses an algorithm to assign each defendant a risk score. A judge then uses that score to decide whether it's too risky to let the defendant wait for trial at home.

An algorithm is a mathematical procedure that uses data to help people make decisions. Algorithms are everywhere. They are best for solving problems that take lots of time if you tackle them one by one but that can, as a group, be standardized. Algorithms are a lot like recipes. When you bake your favorite cornbread, you want it to taste great each time. You want it moist and crumbly on the inside, and golden and crispy on the outside. To make sure you always get the cornbread you want, you probably don't reinvent a new way each time you make it. That would be time-consuming and unreliable. Instead, you probably use a recipe, or mathematically speaking, an algorithm. Your cornbread recipe was probably developed through a long process of data collection. Someone—maybe you, maybe a cookbook author—baked and ate a lot of different kinds of cornbread before landing on the recipe you now use. Similarly, people make algorithms after examining troves of data. The algorithms they make hopefully produce results that are consistent with the data used to make them.

Algorithms make decisions fairer by removing some of the natural mistakes humans make. Of course, preventing human error is more important in solving certain problems than others. If you mistakenly put two cups instead of two teaspoons of salt in your cornbread, the only fallout is an unpleasant taste in your mouth. If you mistakenly judge a defendant as more likely to commit a crime because of your own biases, the fallout could be that an innocent person loses her freedom, livelihood, and chance for a decent life after trial.

The algorithm within the PSA aims to reduce the bias in decision-making about bail. It gives risk scores to defendants based on a calculation done by a computer. The designers of the PSA used a database of over 1.5 million past cases to identify factors that raise a defendant's risk of recidivism. Those factors became a set of questions. The answers that defendants give to those questions are evaluated using a mathematical formula that produces the risk score. If a defendant gets a score of one, the algorithm has determined the defendant is a low risk for committing a crime. If a defendant gets a score of six, the PSA predicts the defendant presents a high risk.

The PSA is one of many "evidence-based risk assessment instruments" in use around the United States. These recidivism tests all use algorithms to determine whether defendants are a potential risk to society before trial. They help courts make high-stakes decisions that affect the lives of defendants and their families, police officers, and the public at large. The entire state of New Jersey now uses the PSA to give judges more information as they weigh whether a person arrested for a nonviolent crime can await trial at home without bail.

The entire state, except sometimes, it seems, for Judge Caposela. Back in the courtroom, the man in the gray uniform sits before Judge Caposela, waiting for the judge's response. The PSA spat out the worst possible number: six, or highest risk. If the algorithm wholly determines his future, his prospects seem grim.

Judge Caposela has the numbers. But he also has his own views of the defendant sitting in front of him. The judge acknowledges the computer's projection. But he insists on drawing upon his own experience in the criminal justice system in making his bail decision. He has seen a lot of accused defendants. He thinks the circumstances in this case warrant overriding the PSA score.

"I'm also sensitive to the fact that he's working and he's supporting four children," Judge Caposela says to the assembled courtroom, as the man in gray and the mother of his children look on. "So, I'm satisfied." Judge Caposela releases the man without bail, on the condition he remains employed. The man collapses in his chair, the tension leaving his body. He wipes tears from his eyes.

Only time will tell whether Judge Caposela made the right call in overruling the PSA. The PSA runs through more data than Judge Caposela ever could. This enormous amount of data and the math that organizes it are supposed to make the PSA and algorithms like it immune from bias. Unlike a computer, however, Judge Caposela is capable of empathy with a fellow human. The judge seemingly allowed his empathetic evaluation of the defendant to control his decision. In doing so, Judge Caposela introduced his own bias by weighing the defendant's family ties over the ruling of the PSA.

The PSA and other algorithms are supposed to eliminate bias. But are algorithms like the PSA actually less immune to bias than Judge Caposela? Algorithms might be made out of math, but they were made by people. People with biases. Sometimes algorithms make decisions that are not fair. Decisions that can have serious consequences for the people affected by them.

For an example of an algorithm wielding its power in potentially biased ways, let's look at two decisions made by a different recidivism algorithm called COMPAS. COMPAS stands for Correctional Offender Management Profiling for Alternative Sanctions, and it was developed by Northpointe, Inc. Like the PSA, COMPAS uses a set of questions, a vast database of past answers, and computers to predict the likelihood a defendant will commit a crime in the future. Its algorithm generates a risk score. A score of one represents "least risky," and ten represents "most risky."

Investigative journalist and mathematician Julia Angwin followed the cases of two defendants whose bail was set with help from COMPAS. The defendants were Brisha Borden, a black teenager, and Vernon Prater, a middle-aged white man. Borden and her friend stole a child's bike and a scooter, property worth about eighty dollars. They rode the vehicles a few blocks down the street and then dropped them, but not before a neighbor called the police. Borden had a record consisting of a few misdemeanors from childhood. But now she was eighteen years old, legally an adult. The COMPAS algorithm gave her a score of eight, labeling her at high risk of committing a crime in the future. The judge in Borden's case told Angwin and her colleagues that he often releases low-level prior offenders with no bond at all for this kind of offense. But he set Borden's bond at one thousand dollars.

Prater was similarly arrested for taking around eighty dollars' worth of property, in his case from a hardware store. Prater also had a criminal record, for armed robbery and attempted armed robbery as an adult. But COMPAS gave him a score of three, thereby deeming him to be a low risk.

How could it be that the COMPAS algorithm gave a high risk score to someone with a prior juvenile record of petty offenses but a low risk score to someone with an adult record involving dangerous felonies?

Given the life-changing nature of decisions made with the help of COMPAS, fairness is critically important and integral to the system's operation. Two examples don't prove that COMPAS is biased or that its bias produces unfair results. But Angwin had a hunch they could be the tip of the iceberg and pointed to a problem that was not isolated to recidivism algorithms. Bias may be endemic to many types of algorithms. So she investigated further.

She began to uncover stories about algorithms that produced results that were significantly biased against minority groups. For instance, Angwin found that the Princeton Review's pricing algorithm tends to charge Asian families higher fees for its test-preparation courses than it charges families of other ethnicities. Angwin also showed that people living in minority neighborhoods in several states often pay higher car insurance premiums than are paid by people who live in predominantly white neighborhoods with the same levels of risk.

These algorithms weren't deliberately designed to discriminate. But that doesn't mean they couldn't be biased anyway. That's because algorithms are built by people. People must make decisions about what to put in an algorithm. This is where bias can be introduced. People who build algorithms have specific goals in mind, which often take the form of minimizing risks and maximizing positive outcomes. Sometimes, the positive outcome for one group is accidentally achieved at the cost of fairness for another group that the designers may have overlooked. It might be a bias of negligent omission rather than deliberate commission.

For example, car insurance algorithms aim to minimize the chance that payments made by drivers do not offset the cost of damage to their cars. Since insurance policies typically cover the consequences of criminal activity as well as driving accidents, the insurances rates will include variables such as differences in car crime rates in drivers' neighborhoods. These algorithms would not take into account a driver's race because there is no correlation between race and driving risk. But using neighborhood crime data may introduce unintentional racial bias into an insurance algorithm because policing practices have historically discriminated against minority neighborhoods.

In COMPAS, bias may similarly be introduced unintentionally through the data that the algorithm uses. The questionnaire that is used with COMPAS incorporates into the algorithm data that weighs against minorities. For example, in trying to assess a defendant's stability and the likelihood that the defendant will remain in the area, the questionnaire asks questions such as, "In the last twelve months before your incarceration, how often did you move?" and "Were you living alone (prior to this incarceration)?" It also asks questions about job prospects, poverty, and optimism about the future, such as "Right now, if you were to get (or have) a good job, how would you rate your chances of being successful?" Measures of poverty, geographic stability, and joblessness are known to correlate with race. Members of minority communities in the United States tend to be poorer, to change jobs, and to move around more than other people. So, an algorithm based on these criteria would negatively affect minorities.

But these measures are also known to correlate with recidivism.

Northpointe argues it needs to include these data in COMPAS in order to produce reliable results. Northpointe insists that it is not deliberately discriminating against minorities but just following the data to data-driven conclusions. Angwin claims, however, that Northpointe could and should have taken into account the ways in which using these measures might introduce bias into COMPAS's calculations. Angwin contends that a further refining of the data could produce an algorithm that is sophisticated enough to take into account the relevant factors without prejudicing minority communities. Doing this would require more data analysis and changes to how COMPAS uses potentially racialized data. It might be hard to do, but it's worth it so that what happened to Borden doesn't happen to more people.

Angwin's suggestions seem worthwhile and workable to fix some of the problems with COMPAS and other recidivism algorithms. But the problems with COMPAS cannot be sufficiently fixed by merely reconfiguring the data. What's wrong with COMPAS, the PSA, and other recidivism algorithms goes much deeper. Angwin's reporting uncovered something much more sinister about COMPAS. Something that gets at the heart of whether it's possible for any algorithm to be satisfactory, let alone fair.

Inspired by Angwin's reporting, three mathematicians, Jon Kleinberg, Sendhil Mullainathan, and Manish Ranghavan, investigated whether it was possible to build a fair recidivism algorithm. They showed that it isn't. Given all the things we care about when we define fairness for COMPAS and the PSA, they showed that we cannot achieve all of them. We want to be fair in too many ways, and some of those ways conflict with each other.

Whether something can be fair seems like more of a question for philosophers than for mathematicians. But just as math helps us design algorithms, math also helps us evaluate their fairness.

Kleinberg, Mullainathan, and Raghavan started by defining all the ways in which we might desire algorithms like COMPAS and the PSA to be fair. They came up with three types of fairness.

All three must be present for the algorithm to be completely fair. But each could be present in an algorithm without the other two, making the algorithm incompletely fair.

The first type of fairness is obvious: the algorithm must make accurate predictions. This means that if the algorithm says a group of people have a forty percent chance of committing a crime before trial, by giving them each a score of four, forty percent of those people must actually commit a crime. Of course, it's hard for an algorithm to do this perfectly. Not even the best algorithm can flawlessly predict the future. But a recidivism algorithm must do a pretty good job to be considered fair.

The second and third types of fairness are somewhat more complicated. Together, they ask that if the algorithm is applied to different groups of people, such as black people and white people, it should not give members of one group higher or lower scores than members of the other without having a good reason for doing so.

In particular, the second type of fairness says that for all people who *do* go on to commit another crime, the average score that the algorithm gives them must be the same across all racial, ethnic, and other social groups. For example, if the algorithm gave all white defendants who committed another crime an average score of eight, the algorithm must do the same for all black defendants who committed another crime. The third type of fairness is almost the opposite of the second. It says that the algorithm must give the same average score to all people in different social groups who *do not* recidivate. For example, if the algorithm gave all white defendants who did not commit a crime an average score of two, it must do the same for black defendants. Taken together, these two types of fairness prevent false positives and false negatives based on social group bias.

Accuracy, no biased false positives, and no biased false negatives. They seem like part and parcel of the same general concern:

the algorithm's predictions should correspond to what happens in real life. Given their complementary nature, you might expect to find many recidivism algorithms that achieve them all.

But Kleinberg, Mullainathan, and Ranghavan found that being fair in one way almost invariably comes at the cost of the other two. This is always true except in two unrealistic and extreme cases. A recidivism algorithm can be completely fair only if you know that all of the members of one group recidivate and all of the members of another group do not, or if you are working with groups of people with nearly equal proportions of recidivists. Kleinberg, Mullainathan, and Ranghavan call the first extreme case "perfect prediction," because group membership perfectly predicts recidivism. They call second "equal base rates," because all groups have the same rates of recidivism. Unfortunately, neither of these cases is real. Black defendants, for example, are more likely to recidivate than their white counterparts. But they do not do so universally. Neither extreme case fits reality. So, perfectly fair recidivism algorithms are mathematically impossible. And the burden of this unfairness almost invariably falls upon members of minority groups, such as black defendants who have done nothing worse than white defendants.

In the paper by Kleinberg, Mullainathan, and Ranghavan, the result appears as Theorem 1.2. It's super mathy, so don't fret if you can't understand it. But I will put it here anyway so that you get a sense of the mathematical nature of the result:

> There is a continuous function f, with $f(x)$ going to 0 as x goes to 0, so that the following holds. For all $\varepsilon < 0$, and any instance of the problem with a risk assignment satisfying the ε-approximate versions of fairness conditions (A), (B), and (C), the instance must satisfy either the $f(\varepsilon)$-approximate version of perfect prediction or the $f(\varepsilon)$-approximate version of equal base rates.

This basically says that the closer an algorithm gets to satisfying all three fairness standards, the closer the groups it works with get to one of the two unrealistic cases, in which there are essentially no social or economic differences among the groups. Put another way, it says that the closer we get to real life, in which race, age, gender, economic status, and other characteristics do not fully predict behavior but in which people of different backgrounds tend to act in statistically measurable different ways, the further we get from perfect fairness.

If we want an algorithm to be fair, it seems that we have to decide which kind of fairness matters most. Do we want the algorithm to be accurate? Certainly, accuracy is important. But then the algorithm will be biased against minority groups. And politicians' main goal for creating recidivism algorithms was to reduce bias.

Do we want the algorithm to produce fewer socially biased false positives? Biased false positives might lead judges to refuse bail to defendants from some social groups but allow bail to defendants from other groups who are accused of the same crimes. Biased false positives might also lead judges to give defendants belonging to certain groups longer sentences than defendants from other groups who committed the same crime. Given these unfair consequences, it might make sense to try to eliminate this sort of bias. But then the algorithm cannot be accurate. And it will lead to mistakes in the opposite direction—letting some groups of defendants off more lightly than is fair, because they actually have a higher risk of recidivism than the algorithm predicted.

This is not a decision that we want to have to make. But as far as the criminal justice system is concerned, it's clear that there are better and worse choices. It is both unfair and unconstitutional to punish people for crimes before trial based on race. So, designers of recidivism algorithms should not make choices that result in the algorithm giving black defendants higher scores than white defendants when they don't recidivate.

According to Angwin, Northpointe made the wrong choice. Its COMPAS algorithm gives defendants of all races who recidivate the same average score. But it is highly inaccurate, and more importantly, it gives defendants who don't recidivate different scores in ways that discriminate against black defendants. This choice has serious consequences for black defendants and for the legitimacy of the system.

Angwin noted that only around sixty percent of those identified by the algorithm as being likely to recidivate actually committed another crime. The algorithm was wrong almost forty percent of the time, violating the first fairness standard of accuracy. Even more troubling, the algorithm's inaccuracies were more likely to turn out badly for black defendants. Almost half of black defendants who were labeled as having a higher risk for recidivism did not go on to commit another crime. But only twenty-three percent of white defendants labeled as such did not recidivate. This is a serious violation of the third type of fairness. Black defendants are more likely to spend more time in jail as a result, before they even have a trial.

It's clear that Northpointe made the wrong choice. But the company didn't see it that way. Northpointe characterized Angwin's finding as at best irrelevant and at worst false because black defendants are more likely to recidivate. While the algorithm singles out black defendants as riskier, Northpointe argues, it does so because they are. While Northpointe may be right that black defendants *are* more likely to recidivate than their white counterparts, Angwin found that the higher rate of false positives among blacks was not related to the higher recidivist rate among blacks. The algorithm, Angwin found, simply singled out black defendants as riskier for no reason other than because they were black.

COMPAS, the PSA, and other recidivism algorithms like them may perpetuate the very problems they were intended to address. The social biases that politicians hoped math could help them avoid turned into mathematical limitations. Those limitations hurt precisely those people the algorithms were meant to protect.

Judge Caposela's deviation from the PSA's calculation makes more sense in the light of this finding. But does this mean we should never use math to help us make high-stakes social and political decisions? Luckily, it does not. It means, however, that we should not let algorithms go unchecked. It also means that when we find biases in mathematical problems, we should do our best to look for unbiased mathematical solutions.

The Bachelor, Algorithms Edition

Before we examine more real-world algorithms, let's take a detour into the world of pure mathematics. COMPAS and the PSA show that when we attempt to use math to help solve social problems, math can be unintentionally wielded in ways that do harm to minorities. Although we tend to think of math as a neutral and objective methodology, it almost invariably harbors subjective bias. And it is not only in real-world algorithms, where you might expect people's social views to affect their mathematical thinking, that biases show up in math. Even a regular math problem, not meant to have any real-world applications, can harbor strange biases.

To get a sense for the pitfalls of designing and evaluating algorithms to address social problems, let's look at an algorithm of the sort that you might use in playing a dating game, choosing a fantasy sports team, or dealing with other situations that require you to make choices based on limited information: the algorithm that solves a popular math problem called the Secretary Problem. While this problem might seem harmless at the outset, widely accepted solutions to it show strangely biased assumptions. Those assumptions are hard to spot. But once you see them, they're harder to forget.

The Secretary Problem goes like this: you're the boss of a company, and you're hiring a new secretary. You received applications for the position. It doesn't really matter how many. It could be ten, or it could be ten thousand. Because we want our solution to this problem to apply to any bosses hiring secretaries, we won't use your specific number of applicants. Instead, we'll just called the number of applicants *A*.

So, you've received *A* applicants for the position, and you're going to interview them. You want to choose the best secretary. Luckily for you, you already know that one of the candidates is definitely better than the others. In fact, you know that the secretaries can be ordered from best to worst, with no two the same. The problem is that while you know that one of them is the best, you don't know which one of them it is. You cannot know until you have interviewed them.

Although you are determined to pick the best possible secretary, you have been saddled with a high-stakes interview process in which you must decide to hire or reject a candidate immediately after you interview the person. And once you reject a candidate, the candidate is gone. There are no call-backs. You will interview them in random order. What's your strategy to pick the best secretary?

What makes this problem tricky, and a tad unrealistic, is that you cannot call back the secretaries after you interview them. We can imagine that maybe the candidates you have rejected immediately go off to be interviewed for another job. Because you must make the call to hire or not in the interview, you can never know the true ranking of the candidates. Unless you interview all of them, you will never know for sure who was the best. But then you'd be stuck with the last candidate, who more likely than not won't be the best.

You may think this problem is impossible to solve. In a sense, it is. It is impossible to develop a strategy that ensures you will *always* hire the best of the applicants. But there is a strategy that turns out the best possible results under the circumstances. As a mathematician would say, you can optimize the situation. The strategy is called the Optimal Stopping Algorithm.

Here's how the Optimal Stopping Algorithm works. The Optimal Stopping Algorithm says you, the boss, should interview the first thirty-seven percent of applicants and hire none of them. All of them go away jobless, no matter how qualified they seem. Then, you should hire the next candidate who is better than all of the candidates who have come before. If you do this, you have a thirty-seven percent chance of hiring the best secretary of the bunch. And this is the best you can do: thirty-seven percent is the optimal probability that the best candidate gets the job.

For example, if a boss had ten applicants (making A equal to ten), the first 3.7 of them would have no chance of getting the job. Of course, 3.7 is an impossible number of candidates to reject. So, since 3.7 is closer to four than it is to three, the boss would actually reject the first four candidates. Then the boss would continue interviewing until a candidate came along who was better than all the previous candidates. If the fifth candidate was better than the first four, that candidate would get the job. If the fifth, sixth, seventh, and eighth candidates were all worse than the first four, but the ninth candidate was better, that one would get the job. And so on.

Using this algorithm, there's a good chance (thirty-seven percent) you will hire the best secretary. In our example, it could happen like this: say the number 1 represents the best of the ten applicants, and the number 10 represents the worst. All the numbers in between represent correspondingly better or worse secretaries. The boss does not know the ranking of the candidates during the interviews, but we will use them to better understand what happens in the example. If the first four candidates would have been ranked 5, 8, 4, and 6,

5 8 4 6,

and the fifth candidate would have been ranked 1,

5 8 4 6 1,

the boss would hire the best secretary. According to the algorithm, applicant number 1 is hired thirty-seven percent of the time. The boss would not, of course, know that this was the best secretary since the boss had not interviewed all of the candidates. All the boss would know is that the hired secretary was better than all of the candidates previously interviewed. We know the boss got the best one because we know the complete rankings. But this should be good enough for the boss.

But there's also a good chance the boss won't hire the best secretary. If the first four candidates were again 5, 8, 4, and 6, and the fifth and sixth were 10 and 3,

5 8 4 6 10 3,

the algorithm would force the boss to hire candidate 3, the third-best secretary. Maybe the boss would have interviewed candidate 1, the best, seventh. But the boss cannot know that this candidate is not candidate 1. The boss could not anticipate that candidate 1 was still to come. In any case, it doesn't matter. Candidate 1 will never see the boss. The algorithm gives candidate 3 the job. At least the boss got one of the better candidates for the position.

Even though you are not guaranteed the best secretary, the algorithm does ensure the bottom thirty-seven percent will not get the job. One of the worst things that can happen to a boss is that candidates 10, 9, 8, and 7 all show up first, and then candidate 6 shows up. The sixth-best secretary would have a lucky day and get the job, despite not even being in the top half of candidates.

But this is not the worst thing that can happen. The absolute worst scenario produced by the Optimal Stopping Algorithm is that you hire no one at all. And this can happen easily, if the best candidate chances to show up for the first thirty-seven percent of interviews.

Here's how this might happen in our scenario with ten candidates. Say the first four candidates were 4, 7, 9, and 1,

4 7 9 1.

All of these candidates would be automatically rejected, because that's the rule. So, bummer, the boss won't hire the best candidate.

But it's worse. Then the rest show up.

4 7 9 1 2 6 8 3 10 5

Because the best candidate showed up in the first batch—the rejects—none of the subsequent candidates is better than those who came before. So, the boss cannot hire anyone. All ten candidates go away without a job, and the boss goes home without a secretary. Even though the second-best candidate arrived fifth, number 2 cannot get the job. It's too bad, because candidate 2 would probably make a great secretary.

And guess what? There's a thirty-seven percent chance that this happens! There's a thirty-seven percent chance the boss will do ten, twenty, eighty-five thousand interviews and hire absolutely no one. All that work for nothing. If you want a thirty-seven percent chance of hiring the best possible secretary, you must be willing to accept an equal chance that the whole hiring process will be a big waste of time.

Now, as the boss, you must ask yourself two important questions. First, is this worth it? Are you willing to exchange a rather high probability you won't hire anyone for an equally high probability you will end the day with the best candidate? The answer depends, of course, on your circumstances. How badly do you need a secretary right away? How badly do you need the *best* secretary? Can you change your hiring practices to include call-backs? This, frankly, is the best option. All the other bosses do it that way.

But what if you're in a situation in which the stakes for picking the best candidate are even higher than just picking a secretary and call-backs are not realistically possible? That is the situation with choosing someone to marry. One popular variation of the Secretary Problem is the Marriage Problem. In fact, when it was first posed by mathematicians, the problem described above actually may have been about selecting a spouse, not a secretary. Some mathematicians muse that the famous astronomer and mathematician Johannes Kepler used a method similar to the Optimal Stopping Algorithm to choose his second wife after his first wife died of cholera. Since Kepler had not been happy with his first wife, he may have wanted to find a more reliable method of choosing his second.

The Marriage Problem and the Secretary Problem are basically the same, except that now you're looking for someone to marry, not a new secretary. In some ways, the Optimal Stopping Algorithm, with its prohibition on call-backs, works more realistically with the Marriage Problem than the Secretary Problem. If you look at dating as an interview for marriage, then you probably would only "interview" one person at a time and decide whether to marry or dump the person. Once you dump someone, there's usually no calling them back for a second round. And it probably matters to you that you marry the best person. Marrying someone who is incompatible would likely result in much more suffering than selecting a less-than-optimal secretary.

In this case, however, a thirty-seven percent chance of finding the best person to marry—equaling a sixty-three percent chance of *not* finding that person and possibly marrying someone worse—

might be too low for you. Alternatively, a thirty-seven percent chance of not marrying anyone might be too high. The Optimal Stopping Algorithm assumes this is a fine price to pay for what you get. You might reject the Optimal Stopping Algorithm on these criteria alone and look for another algorithm to help you solve your problem.

In deciding to use the Optimal Stopping Algorithm to choose your marriage partner, you have to be aware of something else that could confound your plans. The algorithm completely ignores that the person you're dating might have a say in whether the relationship continues or ends. The algorithm assumes that everyone you're dating wants to marry you. You're always the dumper, never the dumpee. Maybe this is true if you're George Clooney. But the rest of us get dumped now and then. Realistically, if you're using the Optimal Stopping Algorithm, it's fair to assume that the people you're dating are using it, too, at the very least because you're probably attracted to geeky, hyper-organized types like yourself.

That many people might find these to be major flaws in the algorithm that render it unacceptable leads us to the second important question: How and why did those who designed the Optimal Stopping Algorithm overlook these drawbacks in the algorithm when you try to use it for important decisions such as hiring secretaries and choosing spouses?

The simple answer is that the question is irrelevant, because there probably isn't anyone who actually uses this algorithm to make those decisions. Apologies to the geeky, hyper-organized crowd, but you're in the minority. The Marriage and Secretary Problems are parables told to contextualize a problem that might otherwise sound pretty dull. When mathematician Martin Gardner wrote about this problem in his column in *Scientific American* in 1960, it sounded nothing like the quaint marriage and secretary stories. He wrote:

> Ask someone to take as many slips of paper as he pleases, and on each slip write a different positive number. The numbers may range from small fractions of 1 to a number the size of a gogol (1 followed by a hundred 0's) or even larger. These slips are turned face down and shuffled over the top of a table. One at a time you turn the slips face up. The aim is to stop turning when you come to the number that you guess to be the largest of the series. You cannot go back and pick up a previously turned slip. If you turn over all slips, then of course you must pick the last one turned.

While it's hard to see, this problem is analogous to both the Secretary and Marriage Problems. It doesn't sound like much fun, though, the way Gardner describes it. Calling this the "game of gogol" doesn't make it any more fun. As much as I would like to spend time with Martin Gardner, who is a math hero of mine, flipping at least thirty-seven percent of a gogol slips of paper in search of a very large number sounds tedious.

But these parables about choosing secretaries and marriage partners teach us about more than just what an algorithm does and what optimization means. They also teach us about how important and difficult it is to account for all the criteria you care about when designing an algorithm that you hope will automate complex, important decisions. Accounting for all the criteria is important because ignoring key criteria can put you in an unfortunate situation. Remember, if you use the Optimal Stopping Algorithm to choose your marriage partner, you have a sixty-three percent chance of not spending your life with a perfect partner, or a thirty-seven percent chance of marrying no one at all. It is difficult to develop useful algorithms partly because what we attend to and ignore can stem from biases, some of which we may not be aware of.

In developing the Marriage and Secretary Problems as examples of the Optimal Stopping Algorithm, the designers incorporated gender and social-class biases. For instance, why did the designers of the Marriage Problem overlook the fact that *two* people, not one, decide whether or not to get married? Could it be because, historically, in the conventional male-centered universe of gender relations, the man proposes to the woman and the woman, swept off her feet, accepts? Could sexist biases be involved here?

This may sound like a stretch. But both of the popular examples of the Optimal Stopping Algorithm, those involving spouses and secretaries, highlight roles in which women were traditionally expected to be subservient to men. And the list of mathematicians who worked on these two problems when they were popular in the 1950s and '60s reads like the list of Supreme Court justices from the same time period—in other words, all were men. So, gender bias is worth considering.

Perhaps lingering biases against the idea that women are able to advocate for themselves sneaked in. And stuck around. Even a different formulation of the problem examined in a paper from the 1980s shows strange biases against women. In this new version, two companies compete to hire the secretaries. But rather than presuming that the secretaries might prefer one company over the other, the formulation has them choose at random between the two companies. No one looking for a job would ever do that. To make things worse, the paper refers to the secretaries as "girls."

The designers of the Optimal Stopping Algorithm most likely did not intend to assume that secretaries and women in heterosexual relationships were passive or that secretaries were naive. But algorithms are used to aid human decision-making. And human decision-making often bends to tacit biases. Given that gender and social-class biases crop up even in such low-stakes situations as solving fictional problems about secretaries and spouses, it is reasonable to assume they do their dirty work on algorithms used in the real world, too.

And it's much harder to laugh them off when they're determining who walks without bail and who stays in jail than when they're just making a math problem into a silly story in order to get out of flipping too many numbers in a boring game with Martin Gardner.

Salamanders, Earmuffs, and Red and Blue Maps

It's also hard to laugh off biases in algorithms when they are being used to manipulate our voting system. This happens all too often through a process with a deceptively silly name: gerrymandering. Gerrymandering is the process by which politicians design the boundaries of voting districts in order to unfairly overrepresent or underrepresent groups of people for their own political gain. That is, based on demographic data about how people tend to vote, the district lines are drawn to ensure the election of members of one political party or a particular racial or ethnic group.

Gerrymandering isn't new. The term comes from a voting district made in the early 1800s by Governor Eldridge *Gerry* of Massachusetts. Governor Gerry had redrawn Massachusetts's state senate districts to favor his political party. This resulted in some crazy-looking districts. One of the districts he created was said to look like a sala*mander*. Critics of what he had done created the facetious word gerrymander to describe in mocking terms what they considered to be cheating by the governor. The term has stuck as a name for the process of manipulating voting districts for some unfair advantage.

Gerrymandering has been criticized over the past two hundred years as unfair and undemocratic. But it is still a widespread problem in our political system today and will continue to trouble us for the foreseeable future. In the summer of 2019, the US Supreme Court ruled that federal courts cannot hear partisan gerrymandering cases, despite acknowledging that partisan gerrymandering is detrimental to democracy. The plaintiffs who brought the case claimed that partisan gerrymandering is unconstitutional because it deprives voters of the right to have their votes count in a meaningful way. They also claimed that it is undemocratic because it can deprive a minority party of ever becoming the majority. And it can allow a party that receives a minority of the votes to get a majority of the legislative seats—potentially forever. The justices—seemingly even those who agreed with the majority opinion—acknowledged the plaintiffs' concerns. Nonetheless, a bare majority of justices ruled that the Supreme Court has no power to prevent these damages. Chief Justice John G. Roberts Jr., who wrote the majority opinion, sent the issue of gerrymandering back to the state and federal legislatures—the very legislatures teeming with politicians who benefit from gerrymandering and would be sorry to see it go.

Whether you think it's a feature or a flaw, in many US states, partisan elected officials get to decide on the boundaries of the districts that elect them. Racial gerrymandering has previously been found to be unconstitutional. Courts have frequently invalidated district maps that were designed to favor white voters over black voters. Gerrymandering against racial, religious, and ethnic minorities is clearly illegal. But whether gerrymandering against opposing political parties in favor of your own is legal or illegal is still in a state of flux. Governor Gerry's original intention has survived to the present day.

Evidence of the unfair and undemocratic effects of partisan gerrymandering is widespread. In 2014, for instance, almost half of the voters in Pennsylvania voted for a Democrat. But only five of Pennsylvania's eighteen districts elected a Democratic representative. This is not the sort of imbalanced representation that most people would expect from a state with fairly drawn voting districts.

Gerrymandering is drawing voting district lines to favor one party. That's unfair. So, the solution would seem to be to just draw district lines that don't favor one party. It seems like a simple problem: find imbalanced representation, and you'll find gerrymandered districts. Straightforward, right?

But it isn't. That's because drawing voting districts that everyone would consider fair is a complicated problem—a complicated mathematical problem. It turns out that it is easier to draw gerrymandered lines than to draw lines that everyone agrees are fair. Politicians hire mathematicians to help them draw districts. Some of the mathematicians are asked to draw gerrymandered lines. Others are asked to draw fair lines. In either case, those mathematicians build—you guessed it—algorithms.

Because algorithms are so complicated, proving that an algorithm introduces partisan bias involves much more than showing that the districts it produces are imbalanced. Just because a district favors one party over another doesn't necessarily mean it is a product of gerrymandering. As we saw with respect to bail algorithms, the outcome of an ostensibly objective algorithm could still be biased. The same is true for algorithms that draw districts: an ostensibly objective district-drawing algorithm could still produce an unbalanced district map. And the legal burden of proof for plaintiffs who are trying to prove gerrymandering is heavy. Those fighting against gerrymandering have to mathematically prove that the existing district is unfair not because of the inherent complexities of algorithms, but for some more sinister reasons. It takes an algorithm to build a voting district, and it takes an algorithm to expose that the district was built unfairly.

POLITICIANS REDRAW voting districts all the time. They've been doing it for over two hundred years. After all this time, the process should be straightforward. But it isn't. Let's look at North Carolina as an example. North Carolina has some of the most strangely shaped voting districts. In fact, in 2017 and 2018, federal courts ruled that North Carolina's districts were unconstitutional. Some were partisan gerrymanders, others were racial gerrymanders. But it seems that their creation started rather innocently, much like all other redistricting decisions do: after a decennial census.

Every ten years, the federal government takes a census. The federal and state governments use information from the census to redraw voting districts. Inevitably, people move around during these ten-year periods. In 2010, North Carolina found, as did many other states, that population density had shifted, mostly from rural areas to urban areas. Every person's vote in a state must count equally with every other person's vote, so all voting districts in a state must have equal populations. So, changes in population density require states to rearrange district boundaries. Drawing districts to meet this requirement is essentially just a division problem, using math you probably learned in third grade.

How might you make voting districts if you were a third grader? Here's one option: simply divide the number of people in the state by the number of districts. Once you know how many people should live in each district, you can split up the state using a procedure I came up with in thirty seconds that I call the Vertical District-Slicing Algorithm. Starting on the state's eastern border, you draw a vertical line through the state every time you've sliced off a strip with the correct number of people living there. Keep going until you've sliced up the whole state. Or, better yet, get a third grader to do it for you while you take a nap. She'll probably enjoy the challenge, and you'll enjoy the nap.

This procedure is an algorithm. It's a simple one—much simpler than the PSA. But it has all the features of an algorithm. It takes a complex problem requiring a uniform solution and uses data to standardize it. It also makes assumptions along the way. And some of those assumptions might be unfair and could end up hurting significant portions of the population.

Possibly the biggest problem with my algorithm is that it assumes that in making district lines fair, all that matters is that we make voting districts equal in population size. But anyone who lives in a representative democracy such as the United States, with its multitude of social, racial, and economic groups that are dispersed unevenly across urban and rural areas, knows that dividing the country into voting districts in this way will not create fair districts. In a democracy, we generally require that all of the significant groups in society be represented in the government. The Vertical District-Slicing Algorithm would always result in districts with equal population size. But the districts would sometimes be long and skinny. They would arbitrarily assign portions of social groups to different districts, and those groups might never have enough voters to elect someone who represents their interests.

Say that several districts included a slice of a large city and a slice of a rural area as they passed through a particularly tall part of the state. With the large city and the rural area sliced up into different elongated districts, neither the city dwellers nor the rural folks would likely get fair representation in the legislature. This lack of geographic compactness in the districts could cause representation problems.

Imagine how California would look if its politicians decided to use this algorithm. California is a long state. It is densely populated in some places and sparsely populated in others. Is it fair to split up the citizens of rural, agricultural central California and group them into voting districts with residents of Los Angeles far to the south? Would such districts elect politicians who adequately represented the people who lived in them? Central California farmers have different needs than urban Los Angelenos. The different politicians they choose to represent them reflect those needs. But grouped into such skinny vertical districts, the city dwellers would routinely outnumber the farmers. The farmers' politicians would lose elections, and their voices would fade.

When we think about representation, the silencing of one group because it is routinely outnumbered owing to weird districting choices is considered unfair. But the Vertical District-Slicing Algorithm is not concerned with whether it disenfranchises farmers. In fact, the algorithm cares about only one thing: district population size. It ignores other factors that reflect the diverse racial, ethnic, and political voices in the United States. The algorithm is on its face perfectly neutral and objective. But its adoption would represent a bias against all sorts of minority groups and interests. The algorithm would create more problems than it solves.

The Vertical District-Slicing Algorithm and the central California farmers it silences may be fictional. But the complex problem of making fair voting districts is not. How can we make voting districts that are fair in all the ways we care about? We want our districts to have equal population. We want them to be geographically compact. We also want them to represent our diverse voices. And we want their creation to be impervious to politicians who might want to silence certain groups, whether for racial or political reasons, or simply because those politicians want to keep their jobs.

Clearly, we need an algorithm more complex than the Vertical District-Slicing Algorithm to draw new districts. But once you begin to care about additional demographic factors, your algorithm rapidly grows in complexity—especially when you're dealing with large numbers of people. For example, in 2018, almost ten and a half million people lived in North Carolina. That's a lot of people. Each of them has a unique demographic profile. It's hard to imagine taking each of them into account when making districts. It makes sense to find a larger, but still representative, unit than individual voters to split into districts. We must look at the different demographic groups that people fall into. Then we must decide which of these groupings are most important for electoral purposes and try to compose fair districts on that basis. Not a simple task, even using high-powered algorithms.

Let's start with a grouping that the government regularly uses to keep track of demographic data: census block groups. A census block group is an area for which the federal government creates a demographic profile. If we break census block groups into voting districts in North Carolina, we only have to break up 6,155 groups.

North Carolina has twelve congressional voting districts. Why don't we build an algorithm that makes all of the possible sets of twelve districts using the 6,155 block groups, compares them along fairness criteria we set, and chooses the best? Then we could be sure we chose the fairest configuration of districts.

But that is easier said than done. There are at least two major problems with this method. The first is a technical one. Taking 6,155 groups and distributing them into every different possible combination of twelve groupings—or "set partitioning," as mathematicians call it—isn't easy. The number of ways to partition a set grows astronomically with the number of items in the set and the number of partitions. There are already 9,330 different ways to break a mere ten census block groups into three districts. If you only had fifty-five different census block groups to break into six districts, you would have to check 8.7×10^{39} different maps. That's more than the number of stars in the universe. It would take anyone, even a computer, an incredibly long time to compare all of these maps. And that's starting with only fifty-five census block groups. North Carolina has 6,100 more. So, this method seems technically unfeasible.

The second problem is both political and ethical. Even if you were able to come up with all of the combinations, on what basis would you decide which was the fairest? You would have to give greater weight to some factors over others. Which ones?

Building an algorithm that draws uncontestably fair voting districts seems impossible. In order to build such an algorithm, we have to be able to answer politically loaded questions about fairness. Then we would have to match our mathematical computing power to the complex answers to those questions. How can you complain about the unfairness of districts that you argue are gerrymandered when you cannot come up with a method of drawing district lines that is uncontestably fair? Arguments about whether a district was gerrymandered are both common and difficult to arbitrate.

For a window into this challenge, let's look at the creation of and debate around one particular district: the Fourth Congressional District in Illinois.

ILLINOIS'S FOURTH CONGRESSIONAL DISTRICT is also referred to by a more colorful name. People in the know call it the "Earmuff District." That's because if you tilt your head to one side, it kind of looks like a pair of earmuffs. In this district, two chunks of Chicago ear-muff another equally large chunk of Chicago that is not part of the district. (Think of this chunk as the head.) The two muffs are connected by a narrow strip of highway. Without the highway, the two halves would be geographically disconnected, a no-no for voting districts. It is a cardinal principle of districting that all parts of the district must touch each other, officially known as being geographically compact.

The Fourth District was designed by Democrats, and it regularly votes Democratic. Republicans object to the design of district. The Republican opponents of the district who sued over it in 2011 argued that the highway does not do nearly enough to make the Fourth District geographically compact. The Earmuff District, they argue, was gerrymandered.

Surprisingly, though, the Democrats who designed the district almost agree. They don't agree that the district was gerrymandered to favor the Democrats. But they acknowledge they drew it to favor a particular group of voters. The Earmuff District has its bizarre shape because the Democrats wanted to keep together a block of Latino residents in Chicago, who also tend to vote Democratic. They claim that the district was designed to allow a minority ethnic group that has historically been subject to ethnic prejudice to have some representation in Congress.

It is an accepted practice to draw district lines so that minority social groups are guaranteed representation. In fact, it's required by law. In 1965, the US Congress passed the Voting Rights Act. It was intended to combat heinous forms of voter discrimination by race that had existed since the Civil War. The Voting Rights Act prohibits politicians from diluting the voting power of minority groups whose members live near each other, tend to vote in similar ways, and have historically faced discrimination by spreading them out over several districts. Spreading them out into different districts would minimize their influence in any single district and make a politician from a minority background unlikely to win an election. The Voting Rights Act says it's a good thing to make voting districts that amplify the voices of historically disenfranchised racial and ethnic groups.

It isn't clear, however, whether drawing districts to give a political party a leg up is legal or illegal. It certainly sounds unfair. Why should politicians be able to effectively rig elections for themselves when drawing voting districts? Nonetheless, federal courts cannot rule against districts accused of being partisan gerrymanders. And politicians are often brazen about making districts that give them an advantage. For instance, the Republican State Leadership Committee, which represents Republicans from all fifty states, started an initiative in 2010 called REDMAP. The goal of REDMAP, which is short for the REDistricting Majority Project, was to support Republicans running for office in states where redistricting would occur the following year. REDMAP wanted Republicans in power so they could control redistricting after the 2010 census and draw districts that would be more likely to vote Republican. This sounds a lot like partisan gerrymandering.

And by REDMAP's own account, they achieved their goal. REDMAP wrote a report titled "How a Strategy of Targeting State Legislative Races in 2010 Led to a Republican U.S. House Majority in 2013." In it, REDMAP bragged that even though "voters pulled the lever for Republicans only 49 percent of the time in congressional races" in 2012, Republicans gained a thirty-three-seat majority in the US House of Representatives. REDMAP managed to draw districts at the state level that give Republicans a national governing advantage, even when a majority of voters preferred Democrats. If this doesn't sound like an admission of partisan gerrymandering, it's unclear what does.

The story of the making of the Earmuff District is similar to the REDMAP story in that the politicians who drew it admitted to having partisan advantage in mind, albeit on a much smaller scale than REDMAP and with the excuse of also trying to empower Latino voters. The Earmuff District was created twenty years ago by Kim Brace, a professional district drawer who was hired by Democrats to create a district to elect a Latino representative. Brace has been hired by representatives across the country to create districts. He takes many factors into consideration when he draws his lines. All of those factors, he and his employers would argue, are legal.

The request that the district help elect a Democratic Latino representative wasn't all Brace had to consider when drawing his lines. He also had to think about keeping the district roughly equal in size to other districts in population. He had to consider whether keeping one ethnic group together might tear another protected group apart. Neither Brace nor anyone else could handle all of this complexity alone. If you imagine Brace sitting at his kitchen table with a map of Illinois sprawled in front of him, drawing and erasing district lines while he adds numbers on his calculator, you're pretty out of date. To draw districts, Brace uses a computer armed with an algorithm. The algorithm takes all of the legally mandated rules of district-making into consideration and, somehow, spits out a map that both satisfies those rules and gives Brace's clients their partisan advantage.

But while algorithms make Brace's job easier, they can make the job of someone contesting a district a lot harder. The Republican-led lawsuit against the Earmuff District argued that the district had been gerrymandered. A better name for district manipulation in this case might be "Quinnmuffing," in honor of the Illinois governor who approved the district, Pat Quinn, and in recognition of the district's infamous earmuff shape.

In their lawsuit, Illinois Republicans argued that the bizarre boundaries of the Fourth District intentionally discriminated against Republican *and* Latino voters. Brace had drawn the district with Latino voters in mind, though arguably not in a discriminatory sense. But sometimes districts that were supposedly drawn to enfranchise a minority group get shot down for disenfranchising that group. For instance, in 2017, the Supreme Court ruled against North Carolina's Twelfth Congressional District for this very reason. The Twelfth District snaked along Interstate 85 to catch three of North Carolina's largest and most predominantly African American cities, Winston-Salem, Greensboro, and Charlotte. The Republicans who drew it called it a majority-minority district, the term for districts that give minority voters a majority as required by the Voting Rights Act. But the Supreme Court called it "packing." Packing is an illegal move that intentionally reduces a voting group's influence across the state by making populations too compact. By packing all of the black voters into one district, the Republicans had limited them from possibility electing more than one black congressperson, who would likely be a Democrat. So, the Supreme Court ruled it unconstitutional racial gerrymandering.

A federal court, however, ruled that Chicago's Earmuff District did not discriminate against Latinos. The court argued that it had kept its weird shape over twenty years "to preserve existing district boundaries and to maintain communities of interest." In other words, the bizarre boundary wasn't packing. It made the district a legal majority-minority district.

Does the earmuff shape intentionally reduce the voting power of Republicans by spreading them out? It may give Democrats an edge, the court said. But it might also be the best of all possible districts. And here is where the algorithm that makes Brace's job easier makes it correspondingly more difficult to identify gerrymandering. The court claimed there were simply too many factors involved in the creation of the Earmuff District to know whether it was a partisan gerrymander. A computer can handle all of those factors, but human judges cannot. There was no way to know whether Brace and the Democrats who hired him could have achieved their legal goals of enfranchising the Latino community without inadvertently hurting Republicans. So many factors went into the algorithm that drew the district that it's too difficult to see if something better was possible.

But, you might wonder, what about checking for imbalanced representation? Couldn't the court determine whether the Earmuff District gave Democrats an unfair advantage by looking at election results? Researchers could find the number of registered Democrats and Republicans in the state. If there are more Republicans, but Democrats seem to be winning most of the elections, you might have a good argument for partisan gerrymandering.

Even if this were the case—which it wasn't in Illinois—imbalanced representation isn't by itself proof of partisan gerrymandering. This is partly because drawing districts that give minority groups fair advantages do sometimes incidentally—even if not accidentally—privilege one party over another. The best possible district for Chicago Latinos might also give Democrats a bit of a leg up. Under the law, this is considered a fair trade-off if we want to overcome historic voter discrimination.

Also, sometimes imbalanced representation happens for reasons outside of the control of people drawing districts. For example, Democrats tend to live in cities. Republicans tend to live in rural and suburban areas. Most experts on district-drawing agree that districts should be geographically compact. But this means that city-dwelling Democrats tend to be packed into a few districts, while country-living Republicans are spread out enough to get majorities in more districts than their proportion of a state's population would seem to entitle them.

Furthermore, imbalanced election results can stem from a baffling but true fact about close elections in our district-based system. When the divide between Democratic and Republican voters in a state is close to fifty-fifty, it is likely that elections across districts won't get close to fifty-fifty results. This is because it only takes a few votes to swing each district from one party to another. In states where the partisan divide is tight, those few votes could easily swing many more than half of the districts to one party or the other. Ironically, it's in states where you would least expect one party to dominate that you often find the most imbalanced election results.

An imbalanced result might feel unfair. But it might have happened for reasons politicians couldn't control. In any case, unfairness isn't inherently unconstitutional. That's what judges have said, at least. They say that whether they are cases of illegal gerrymandering is just too complicated for us to know.

Sometimes the biases behind decisions stand out like neon signs in the dead of night. At other times, people, including Supreme Court justices, disagree about those biases enough to debate whether or not they exist. Judges are not supposed to interfere in political processes and decisions unless they have clear grounds for doing so and a clear remedy for the problem. As a result, judges usually punt when confronted with cases as complex as partisan gerrymandering.

In 2004, Justice Antonin Scalia wrote an opinion for a Supreme Court case about partisan gerrymandering. He wrote that "no judicially discernable and manageable standards for adjudicating political gerrymandering claims have emerged. Lacking them, we must conclude that political gerrymandering claims are nonjusticiable." Based on the green light that Justice Scalia seemed to be giving them, some politicians made it clear they would redraw districts to benefit their political parties. Some even explicitly declared they were considering only the advantage they were giving to their political party in drawing districts so they would not be accused of drawing racist districts. For example, North Carolina politicians went on the record that they were making districts to benefit Republicans, not white voters, after their districts were ruled unconstitutional for disenfranchising black (and often Democratic) voters. North Carolina Republican State Representative David Lewis said, "We want to make clear that, we—to the extent, are going to use political data in drawing this map, it is to gain partisan advantage on the map. I want that criteria to be clearly stated and understood." No need to hide it if it's nonjusticiable.

But all may not be lost in trying to rid our political system of gerrymandering. That's because mathematicians have made a lot of progress since Justice Scalia wrote his opinion in 2004. Mathematicians, it turns out, are particularly good at juggling variables. We actually can manage this complexity, mathematicians say. We can manage it to such an extent that we might be able to attribute imbalanced voting results to intentional partisan gerrymandering and make partisan gerrymandering justiciable. What we need is another algorithm.

ALGORITHMS CAN BE USED to perpetuate biases. But in the hands of socially aware, conscientious mathematicians like Wendy Tam Cho, they can also be used to uncover biases. Cho claims to have built an algorithm that will reliably find partisan gerrymandering and provide fairer alternatives, even in the most complex situations.

Cho says she has always been fascinated by power. Her mathematical work is driven by the troubling question of, "How is it that in a human society, we can organize ourselves into governance structures so that . . . some people have power and other people do not have power?" People can wield mathematics to unfairly distribute power. But Cho takes mathematics back. With algorithms, she gives the power back to the people.

To recap, here's the problem mathematicians face when trying to build an algorithm that finds gerrymandered districts and constructs fair districts: they want to build an algorithm that will draw all possible legal districts and see which one is the fairest. But they can't do this because the number of possible districts is astronomically huge. Remember, North Carolina has 12 districts and 6,155 census block groups. Even a supercomputer cannot create all possible districts in a reasonable amount of time, let alone analyze which one works best.

Cho's solution to this problem sounds relatively straightforward: If you can't check all of the districts, why don't you check a smaller sample? But figuring out which sample to check is mathematically complicated. You could just choose a smaller sample of possible districts at random. But the random pool might not be useful, because many randomly drawn districts aren't realistic. Alternatively, you could narrow the list of criteria you care about when drawing the districts. This would also produce a shorter list of districts. But we still need the short list to reflect the American demographic landscape. Any criterion we remove will make our district-generating and district-comparing algorithm less accurate and relevant. But any criterion we add will make it more difficult for an algorithm that selects districts at random to cover the space of districts and choose a representative sample.

Cho and her coauthor, Yan Liu, knew they somehow needed to narrow the list of districts they checked for fairness. But with random sampling and shortening the list of criteria ruled out, what could they do?

Cho and Liu came up with a better method. They developed an algorithm that draws what they call "reasonably imperfect plans." These plans satisfy legal requirements and aren't gerrymandered.

They also meet criteria particular to the political landscape, making them feasible for governments to implement. By narrowing the range to only "reasonably imperfect plans," Cho and Liu weeded out some of the more outlandish possibilities and gave themselves a more manageable set of plans to check. A supercomputer uses the algorithm created by Cho and Liu to build the plans. Now that they have a list of reasonable plans, Cho and Liu select districts at random from among this smaller list.

Their randomly selected plans then face the final test: Are they more or less fair than a district that politicians claim is gerrymandered? Cho and Liu can evaluate whether the contested district does worse, better, or about the same as other districts with respect to the criteria people are fighting about, such as favoring one political party or racial group over another. If the contested district performs just as well as the simulated districts on treating political or racial groups equally, it probably wasn't gerrymandered. But if it performs worse, Cho and Liu have mathematical evidence to support an argument that it was gerrymandered. If many better districts exist in the set of reasonably imperfect plans, perhaps the contested district was drawn for political or racial reasons. And Cho and Liu have overcome Justice Scalia's conclusion that partisan gerrymandering cases were not justiciable because we cannot provide a remedy for the problem.

Cho and Liu used their innovative algorithm on Maryland's voting districts, which Republicans argue unfairly favor Democrats. The algorithm identified about 250,000,000 maps that did at least as good a job of meeting the legal criteria as the map Maryland already had. It then narrowed this massive list down to about 250,000 maps that constituted the set of "reasonably imperfect plans" from which those who were drawing Maryland's districts could reasonably choose.

How did Maryland's map compare to the quarter of a million other viable maps in terms of partisan bias? There are many ways to examine a map for partisan gerrymandering. Cho and Liu chose to look at how the number of seats a particular party won in an election responded to changes in the percentage of voters who favored that party. In a fair system, if the percentage of Democrat voters dropped, one would expect the number of seats won by Democrats to also drop. But with a less responsive map, the number of seats won by Democrats would drop less. The less responsive a map is to changes in voter preferences, the more likely it was gerrymandered.

Before you read the results of Cho and Liu's study, take a moment to set your own personal threshold for Maryland's district. What portion of the 250,000 other possible maps would have to be more responsive to changes in voters' political preferences before you would call Maryland's plan gerrymandered? Would you be strict with the drawers of the map and say one-quarter? Politicians tasked with such an important job as drawing fair voting districts should outperform even a supercomputer, you might argue. Or would you be equitable and say half? Indulgent with seventy-five percent?

Any threshold you set probably will not come close to the actual percentages of simulated maps that Cho and Liu found outperformed Maryland's. Almost ninety-five percent of the districts drawn by the supercomputer were more responsive to changes in voters' political preferences than the map Maryland already had. Or, put another way, Maryland's map is so bad that if politicians chose the map by pulling district maps out of a hat, they'd only have a five percent chance of picking a map as bad as or worse than Maryland's. Those aren't great odds. Cho and Liu's algorithm shows that it's likely Maryland's map is a political gerrymander.

Cho and Liu's algorithm isn't perfect. Critics argue that comparing the responsiveness of contested and reasonably imperfect districts isn't the best way to assess a district. Remember, in states with close elections, representation can be lopsided even when district boundaries aren't gerrymandered. But Cho and Liu's district-simulating algorithm goes a long way toward capturing the complexity of real-world districting problems. Even better, it produces information that people, particularly legislators and justices, can use to determine gerrymandered districts and require that fairer district lines be implemented. It breaks new ground in solving a math problem that many experts feared could not be solved. With mathematics, it tilts the balance of power back toward the people.

Perhaps algorithms have been misused to make our political system unfair. But these powerful tools have promise. We just need to keep checking the work of the people who make them.

4 Can Math Open Doors?

Math and the Problem of **OPPORTUNITY**

"Many People Think That for Math You Have to Be Perfect"

A soft-spoken girl, not more than fifteen years old, sits in front of the camera. Her wide-eyed joy is evident despite her large glasses and the dark hair draped across her eyes. She says,

> I would say that math is art. . . . It can change your mindset, and it can open, like, billions of worlds in your head with formulas, and you can create infinite, like, numbers. And, it's just really fun to play with. And many people think that for math you have to be perfect—you have to do the formula this way, that way. But you can also *explore* with these formulas and with these numbers. And create something so really cool—*really* really cool.

These words may disclose her youth, but they don't reveal that her early school years were, as she puts it, "not so great." That she is Latina and grew up in the Bronx. Or that she has not previously enjoyed math. And yet now she says that math is "fun to play with." And that math allows her to create "something so really cool—*really* really cool."

When we imagine a person who says that math "can open billions of worlds in your head with formulas," do we picture someone like this young woman? And if we don't, why not?

The young woman in this interview would consider herself a novice mathematician. Professional mathematicians from her demographic are rare. The Annual Survey of the Mathematical Sciences found that of the 1,926 math PhDs awarded in 2014, only 13 went to Hispanic or Latina women. Hispanic and Latino men did a little better, earning 54 of those PhDs. Together, Latinx make up no more than four percent of the total math PhDs while making up almost twenty percent of the population of the United States.

Even more relevant for this young woman, given her age, is that only nineteen percent of Latino eighth graders were deemed proficient in math in 2015 on the National Assessment of Educational Progress. In contrast, forty-three percent of white eighth graders reached that important math benchmark. This young woman is one of a very few.

The relative absence of women and people of color in the fields of science, technology, engineering, and math—also known as STEM—is a well-known problem. Of the four STEM fields, math is the least diverse. This is not merely an academic problem. It is also a social problem. Lack of diversity in STEM is the source of other problems in our society. A significant amount of the economic opportunity in our society is related to expertise and academic degrees in STEM fields. Minority groups that are shut out of STEM fields are shut out of a large swath of economic opportunities, which can hurt these groups economically, socially, and politically. It can also leave them out of the loop when it comes to solving problems that may directly affect themselves and their communities. People with degrees in STEM are on the front lines of solving some of the world's most important social problems. When the problem solvers don't adequately represent the people whose problems are being solved, inequities can result. There is perhaps no better illustration of the impact of this problem than the story of the recidivism prediction algorithms that discriminate against black defendants.

Lack of diversity in STEM programs is the source of a vicious cycle of educational problems for girls and students of color. When few women and people of color study math at high levels, the body of teachers in those fields will also lack diversity. When students like the young woman above do not see people similar to themselves excelling in math, they have trouble imagining themselves excelling. Their odds of becoming mathematicians decreases. And so the vicious cycle continues, with women and people of color failing to pursue math at a higher level because they didn't have math teachers who were also women and people of color in school.

So, how can mathematicians help to break this cycle? Many are working to do so. But before answering this question, we must explore a different one: What does it take to become a professional mathematician?

Spoiler alert: doing well in school math classes or on the National Assessment of Educational Progress isn't enough.

TERENCE TAO WAS A CHILD MATH PRODIGY. By the age of nine, he was taking math classes at a university near his childhood home in Adelaide, Australia. At seventeen, he embarked on his doctorate at Princeton University. Tao has won the Fields Medal, one of the most prestigious awards in mathematics, and a MacArthur Fellowship (popularly known as a "Genius Grant"). At age ten, he was the youngest person to medal in the International Mathematical Olympiad. No surprise, then, that Tao is now a math professor at the University of California, Los Angeles (UCLA).

Knowing Tao's mathematical prowess, it seems obvious that he would become a professional mathematician. Such ability must be rare. Some have called Tao a genius among geniuses. How could he not become a mathematician?

If these are your thoughts about Tao and his success, you are in good company. Many prominent mathematicians attribute mathematical ability to genius.

One of the most famous mathematicians of all time, Henri Poincaré, had a similar explanation. Mathematicians, he argued, have an ability that others lack—an ability that makes their choice to become mathematicians almost inevitable. Poincaré calls that ability a "special sensibility." He wrote in one of his books that mathematical discoveries are those "that can charm that special sensibility that all mathematicians know, but of which laymen are so ignorant that they are often tempted to smile at it."

According to Poincaré, these lucky few have a genius that transforms an incomprehensible subject into something that is almost obvious to them. To the mathematician, Poincaré writes, math is a subject "whose elements are harmoniously arranged so that the mind can, without effort, take in the whole without neglecting the details." It's hard for mathematicians to understand why others don't experience math as they do, Poincaré laments with what seems like mock sympathy for the unfortunate masses. Nonetheless, it's impossible for them to communicate their mathematical ideas to the masses of people who do not possess mathematical genius. Poincaré writes,

> One first fact must astonish us, or rather would astonish us if we were not too much accustomed to it. How does it happen that there are people who do not understand mathematics? If the science invokes only the rules of logic, those accepted by all well-formed minds, if its evidence is founded on principles that are common to all men, and that none but a madman would attempt to deny, how does it happen that there are so many people who are entirely impervious to it?

What distinguishes the mathematician, then, according to Poincaré, is this "special sensibility." Poincaré does not address where this sensibility comes from. He makes it clear, however, that few people have it, and those who do not have it never will. According to Poincaré, mathematical ability cannot be learned. Either you were born with it or you must learn to live without it.

This may seem like an extreme position. But it's consistent with what has been the widespread conventional view of mathematicians as geniuses. Genius isn't learned. It's gifted at birth. And it would be easy to conclude that someone like Tao is one of the few who possess this special sensibility. Tao certainly has a talent for math, and this talent accounts for a good deal of his success. If Poincaré is right, then people who become mathematicians are destined to do so.

This position implies some disturbing conclusions about why there aren't more women and people of color in mathematics. Let's follow Poincaré's position to its logical conclusion: if all mathematicians are born geniuses, attracting more women and people of color to math would mean finding those geniuses among them. But if mathematicians have not over many centuries found more geniuses among women and people of color, it would be reasonable to conclude that they aren't out there. Do we believe that there are so few women and people of color in higher levels of mathematics because they simply don't have that genius? Poincaré might say so. And the implications of the conventional view of mathematicians as born geniuses would imply likewise. Must we agree?

I hope not. So, how do we refute Poincaré's claims? One place to start would be to look at the work Tao does. Is it really as incomprehensible to non-mathematicians as Poincaré suggests? Does it take genius to appreciate what Tao does, and maybe to join him in his work?

One of Tao's most important contributions to mathematics is in a field called number theory. He has been trying to solve a problem in number theory that has been unsolved for over two thousand years. The problem is deceptively simple. But it's also tantalizingly elusive.

Much of number theory is like this: simple, yet elusive. Number theory deals with properties of the most basic numbers, the numbers that we use to count. Mathematicians call these counting numbers whole numbers. You probably spent most of your elementary school math years studying whole numbers. Skip-counting, factoring, finding greatest common divisors. These are all topics taught in elementary school. But they are also involved in some of the most interesting problems in number theory. In fact, the unsolved problem that Tao is working on deals with some of the most important numbers to both mathematicians and elementary school students alike: prime numbers.

Here's a brief elementary school recap if you don't remember what primes are: prime numbers are whole numbers that can only be cleanly divided by themselves and one. "Cleanly divided" means that when you divide one whole number by another, you get a third whole number, not a fraction or decimal. Four is not a prime because it can be cleanly divided by two as well as by itself and one. Five, however, is a prime. No matter how hard you try, you will never find any whole numbers other than five and one that divide it without giving you a fraction. The first five primes in the ascending numerical order are two, three, five, seven, and eleven.

Prime numbers are simple enough to be taught to elementary school students. But they are also mysterious enough to be the subjects of ongoing investigations by many of the most prominent mathematicians, including Tao.

The unsolved problem that Tao is working on is called the Twin Primes Conjecture. It goes like this: show that there exist infinitely many prime numbers that are two numbers apart. This sounds simple, but it has gone unsolved since Euclid allegedly posed it over two thousand years ago.

Among the smaller numbers, twin primes are relatively common. Three and five, five and seven, eleven and thirteen, seventeen and nineteen are all twin primes. But they eventually start to thin out. No known pattern predicts how often they occur. As of the end of 2016, $2{,}996{,}863{,}034{,}895 \times 2^{1{,}290{,}000} - 1$ and $2{,}996{,}863{,}034{,}895 \times 2^{1{,}290{,}000} + 1$ were the largest known twin primes. These are big numbers, but not big enough to show that twin primes would continue to appear if we were to continue to count up infinitely.

Tao's contribution to solving the Twin Primes Conjecture was to change the question slightly. Tao and his colleague Ben Green had the idea to look for lists of primes that were some number other than two apart. They thought that these other groups of primes might teach them something useful about twin primes. For instance, Tao and Green looked at what you might call "quadruplet primes," primes that are four apart (three, seven, and eleven), "sextuplet primes," six apart (seven, thirteen, and nineteen), and even "triacontaplet primes," thirty apart (151, 181, 211, 241, and 271). Their hope was that the way these other groups of primes appeared in the ascending numerical sequence might help them see something about twin primes.

It is a reasonable approach. It also doesn't seem like the kind of idea that only a genius would have. If you're struggling to solve the problem at hand, it is understandable that you might try to solve a more accessible but related problem. If you succeed with the related problem, you might be able to apply what you did toward solving your original problem. People take this approach to problem-solving all the time, in and out of math. And you don't have to be a genius to think of doing it. Tao and Green thought they could generate many more primes that were a number other than two apart, which might help them with the Twin Primes

Conjecture. Because they adopted this method, they had much more material to work with for figuring out the way primes group themselves. Even a non-mathematician might understand how this method makes sense.

And it worked. After studying their new lists of primes, Tao and Green made a discovery that puts solving the Twin Primes Conjecture within closer reach. They constructed different groups of primes with the primes in each group separated by a different number of whole numbers. They then compiled and compared lists of these groups of primes. Tao and Green found that no matter how long a list of primes separated by a constant gap, they could always find a longer list, separated by some other constant gap. They still don't know whether twin primes go on forever. But they do know that lists of primes separated by some gap do. The list of primes that are some whole number apart can be as short as two and as long as you like.

You may not understand everything that is going on here if you're not familiar with the terminology and technics involved in this sort of math. But that's okay. The point is that the method Tao and Green adopted in solving their problem was a fairly common strategy. It is used by all sorts of people when solving the most mundane as well as the most advanced problems. Figuring out that lists of primes separated by some gap go on forever certainly took specialized mathematical knowledge that most people do not have. And no mathematical genius, not even Tao, was born knowing all the intricate details of how to work with large prime numbers. They have to learn the specifics.

If you taught the same math specifics to an ordinary person and to a genius, would the ordinary person get it? Would the ordinary person go on to be able to come up with ways of helping solve problems such as the Twin Primes Conjecture? Poincaré would say no. The "special sensibility" that Poincaré suggests all mathematicians have is something more like special insight into how to approach challenging problems. Such as knowing that you often make headway on a difficult problem by tackling a simpler related problem. But is this really a genius-level insight? Or is it more like the sort of insight that many people have about problems in their own fields, and that they learn how to follow as they gain experience and confidence? Poincaré would say it is more like the former. I think it is more like the latter. And the difference between whether you see math as a subject for geniuses or as something that ordinary people can do makes a big difference in the way you approach math education and who you think can be a mathematician.

Saying that Tao and Green's key insight into the Twin Primes Conjecture isn't genius does not diminish the significance of their contribution. No one had thought to pursue their approach before. Not in over two thousand years. Tao and Green have some ability that others lack. Others who make it to the ranks of professional mathematics must, too. How did they get it?

If it isn't, as Poincaré would have it, a matter of genius and that genius will come out irrespective of the circumstances, then what is it? Perhaps it's school. Did Tao, Green, and other mathematicians like them get their start as mathematicians in school?

It's possible. But the math that most children experience in school does not resemble the kind of work that is done by professional mathematicians. It's hard to get a real impression of what it's like to be a mathematician from school. Sure, you worked with some number theory when you were in elementary school. You learned what primes are and how to factor. But did you know that, even after two thousand years, mathematicians *still* don't know if twin primes go on forever? The Twin Primes Conjecture is a mathematical mystery that even elementary school students could understand. It would probably also interest them. Kids like puzzles, and that's what the Twin Primes Conjecture is. And yet most of them probably know nothing about it.

Professional mathematicians grapple with mysteries like the Twin Primes Conjecture every day. But children don't imagine mathematicians as mystery solvers and pattern sniffers. Most children don't know anyone who is interested in math other than their teachers and probably have never met a professional mathematician. How they imagine a mathematician develops from representations of mathematicians in the media. And these representations are often distinctly unflattering.

In one study, education researchers asked five hundred middle school students from five countries to draw a "mathematician at work." A jarringly large number of children drew an adult who was coercing children to do math with violence. Some of the pictured mathematicians even wielded weapons. Not a flattering portrait of what were apparently their math teachers. Many other children drew a foolish person completing simple math problems incorrectly. Seemingly, their image of the mathematician was as a geek, and not a very smart one. In addition to being stupid and violent, a vast majority of the mathematicians the students drew were men. Even the girls mostly drew male mathematicians.

Plenty of professional mathematicians and math teachers do not fit the description that these children illustrated. But most children do not get to meet real mathematicians or deal with the sorts of problems that mathematicians work on. Their ideas of math and mathematicians comes from their schoolteachers, textbooks, and caricatures of geeky mathematicians that they see in the mass media. Most math teachers are not viciously coercive. But because many students find the math they are taught as boring and oppressive, many transfer their feelings about math onto their math teachers. Most professional mathematicians are not nerds, but that is how they are frequently portrayed in popular culture. Filled with images of coercive math teachers and nerdy math professors, who are almost always male and white, what child, and particularly what girl or kid of color, would aspire to a career in mathematics?

Tao probably wouldn't have. At least, that's what he says about his childhood. When he was a kid, Tao says that he had no idea what being a mathematician involved. But Tao got lucky. His parents noticed his ability in math. So, they sought advice from local math professors. Those professors put Tao in touch with a tightly woven community of mathematicians from around the world who congregate and communicate with each other in professional and social organizations, and who try to foster aspiring mathematicians. The mathematical community eagerly embraced young Tao, presenting him with opportunities to experience math beyond what he was given in school. Attending Princeton at age seventeen was the culmination of many years of outreach. By that time, Tao was well connected to many of the mathematicians who helped to shape his work. Without that welcoming embrace from the community of mathematicians, Tao probably wouldn't have made it to where he is today.

Tao isn't the only professional mathematician with this kind of story. When mathematicians talk about their childhoods and educations, the amount of specialized support they received stands out. Mathematicians frequently come from families of mathematicians. For example, mathematician Don Zager's father inspired his love of math. "We'd walk in the woods and he'd stop in the middle of the walk to show me the Pythagorean theorem and to point to it in nature," he said. "He admired mathematics very much and I think it meant a lot to him that I was drawn to it."

Mathematicians also often tell stories of teachers who introduced them to mathematics beyond the standard school curriculum. Zager's math teacher at age eleven made "special rules" for him. "I could read math books during class or work on other problems," he said, "but on tests I would receive a zero grade unless everything was perfect." Knowing Zager's background, it is easier to understand how he achieved a career in mathematics. He had two bachelor's degrees from the Massachusetts Institute of

Technology (MIT) by the time he was seventeen. It's hard to imagine how Zager could have achieved his success without a parent who used math to bond with him and a teacher who nurtured his talents and interests.

The same is true of Tao. In a world where so many students do not choose math, it is worth stating: Tao could not have medaled in the International Mathematics Olympiad at age ten if he did not know about it or aspire to compete in it. It took the proverbial village to bring Tao to the mathematics department at UCLA. His parents sought expert advice when their preadolescent son showed promise and subsequently soared through university math courses. The founder of Center for Talented Youth, a summer program for academically ambitious kids, advised Tao's parents to let their son's interest in mathematics bloom. Elias Stein, Tao's advisor at Princeton, took Tao aside for a pep talk when he performed poorly on an important exam and encouraged him to soldier on. Tao had dozens upon dozens of collaborators, his wife and children, and countless others cheering him on.

Perhaps Tao was born with mathematical ability that others lack. But one can clearly see that Tao, and so many other mathematicians like him, had something at a young age that might matter more than innate ability: an invitation to join a world where math matters.

TAO WAS LUCKY TO RECEIVE an invitation to join the community of mathematicians. Plenty of other budding mathematical minds do not, especially those belonging to girls and students of color.

To become a mathematician, you often have to know a mathematician. You need the support of a community of people who can envision the mathematician that you might become. Having this community is a form of privilege. Ignoring the influence of community on mathematical development by attributing unlikely mathematical success to genius closes pathways for students whose mathematical talent goes unrecognized because no one from the mathematical community spots and nurtures it. To honor these students, we must tell their full story.

An overriding question in mathematics—a genuine math problem—is, How can mathematicians extend their invitations to a broader group of people? How can they reach out in particular to mathematically talented girls and students of color? For example, how can they ensure that more than one percent of the math PhDs awarded in any given year will be earned by Latina women?

These are the questions being taken on by a group of mathematicians and math educators in New York City. They are helping students from backgrounds that are traditionally underrepresented among mathematicians get into selective math summer programs and New York City's elite math and science high schools. Their group is called BEAM, or Bridge to Enter Advanced Mathematics.

BEAM recruits rising eighth graders from New York City's heavily black, Latino, and Asian outer boroughs to spend three weeks at a math summer camp in Upstate New York. It introduces kids to the kinds of mathematics they might only see if they pursued math in college: pure math topics such as number theory, topology, and combinatorics, and applied math topics such as astrophysics and programming. It teaches them the language of proof. And it exposes them to the kind of logical debate that mathematicians engage in when working on a problem.

In New York City alone, BEAM faces some challenging statistics. New York City has over three hundred thousand black students enrolled in public school. That's twenty-six percent of total New York City public school enrollment. But only twenty-four of those black students go to the most selective math and science magnet high school in the city, Stuyvesant High School. Black students make up a mere 0.72 percent of Stuyvesant's student body.

The organizers of BEAM insist that the underrepresentation of black students and students of color generally in New York

City science and math magnet schools is not a talent problem. It's not a lack of geniuses, as Poincaré might put it. There are many more mathematically talented students of color in New York City than are represented in the city's elite public high schools. All of the students who attend the BEAM summer program were recommended by their seventh-grade math teachers as talented and interested math students. They also passed BEAM's rigorous entrance exam. But BEAM recognizes that talent is not enough to get students from nontraditional backgrounds into advanced mathematics. What these students need is the sort of outreach that helped Tao break into the ranks of professional mathematicians. They need an invitation to join the mathematical community.

BEAM founder Dan Zaharopol got an invitation when he was a kid, and it made all the difference in his life. Like Tao, Zaharopol is a math success story. He developed his love of math at a young age and went on to study math at MIT. Young Zaharopol had a lot in common with the students from New York's outer boroughs that he now recruits for his camp. He showed talent for math at a young age, loved puzzles, and often felt bored in regular school math classes. But he also had special advantages that most BEAM students do not have.

Zaharopol's childhood was different from that of his BEAM students in some important ways. Zaharopol is a white man from a well-educated family that had the wherewithal to foster his interest and ability in math. These advantages set him up to achieve in mathematics. In organizing BEAM, Zaharopol knew that he could not simply translate the academic advantages that come with being white, male, and from a college-educated family to black and brown preteens from New York City's poorest neighborhoods. So, Zaharopol took a closer look at what specifically he and others like him had been exposed to at a young age that set them up for mathematical success and translated those experiences into BEAM's program.

Zaharopol came up with three essential opportunities that might be out of reach for some students but that BEAM could offer them. The first was math enrichment opportunities outside of school that Zaharopol had been offered. He went to science and math summer camps, joined math teams that competed in contests around the country, and learned computer programming after school. For him, math was not just a school subject. It was a year-round activity. And it was fun.

The second was collective knowledge of how to find math enrichment opportunities that his family and other families in their social community had. They shared this knowledge after using it to give their kids a leg up. Schools do not always let students know about enrichment opportunities available beyond what the school itself offered. Knowledge of programs like the ones that Zaharopol attended was collected by parents over time and passed from family to family within the community.

The third was the encouragement that Zaharopol received from family, friends, and teachers that enabled him to aspire as a young student to a career in math. He may not have known precisely what to aim for. But from an early age, Zaharopol considered math-related education and jobs to be both plausible and admirable futures for himself.

Providing these three kinds of opportunities to students who might not otherwise have them is the core program of BEAM. Aspiration needs nourishment in school and beyond. But kids cannot find that nourishment without specialized guidance from adults. Without access to the communities of people who work in math or an outlet in which to do fun math activities with peers, kids are unlikely to develop interest in math that grows into focused drive. A drive to pursue math can quickly cool if opportunities seem out of reach. And these three opportunities offered by BEAM are not provided to kids by their schools. They are either provided by the communities they live in or not provided at all.

Zaharopol was fortunate to grow up in a community that did provide such opportunities. BEAM seeks to become such a community for students who might not otherwise have these opportunities. The opportunities available to Zaharopol in essence flowed from his being raised in a culture that included mathematics. Kids like Zaharopol, raised in a community of college-educated adults, see things around them all the time that encourage mathematical aspirations. They are generally no more than a few friends apart from someone with a degree in math or science. Many hands reach out to pull them further into math. Kids from backgrounds underrepresented in math have trouble gaining access to these activities and to encouragement. They are not surrounded by academic opportunities. BEAM's three-pronged approach helps give students that access.

And with access, BEAM students blossom into mathematicians. These kids beat the odds. In subsequent years, they head off to competitive math camps, highly selective math- and science-focused New York City high schools such as Stuyvesant and Brooklyn Technical High School, and prominent universities. For these kids, BEAM is just the invitation they need to join the community of mathematicians.

When an Invitation Isn't Enough

Although BEAM does a great job of welcoming students from backgrounds underrepresented in mathematics into the often tight-knit community of mathematicians, sometimes access isn't enough. Getting an opportunity to join the community does not ensure the opportunity to become a true member.

Even if students of color do make it into an elite school or summer program, being one of a handful of students of color in a field long dominated by white scholars can present numerous hurdles. These students are not only often unfamiliar with the language of higher-level math, which their more privileged peers already may have learned in previous enrichment programs or from family members. They are also generally unfamiliar with the cultural practices of their new community. And their new community is unfamiliar with them.

Being one of a handful students of color in a predominantly white institution can be difficult. Being treated as "special" or uniquely gifted for having broken through can make students of color feel singled out, isolated, and alone. And these students might not just feel isolated. Their white peers and instructors might do things, intentionally or not, to isolate them. Their white peers might not be eager to see new, different faces. The field in which these students are trying to work is not accustomed to responding to their voices.

So, being a student of color in such a setting presents more than just mathematical challenges. And if students of color happen not to make it in the world of math, despite getting that much-lauded invitation, they can end up blaming themselves, rather than considering the circumstances with which they had to struggle. Making it as a novice in an academic community is inherently more difficult for the newcomers than for those raised in that community. Academic institutions, like most other institutions, are structured for the success of insiders, not outsiders. Nonetheless, blame for failing to succeed is most often placed upon and felt by those outsiders who can't make the grade.

For a stark example of a young mathematician of color for whom an invitation to join the mathematical community was not enough, we turn to a story that mathematicians often treat as one of success. Success at diversifying the field. Success at giving a gifted young man from an underrepresented background the leg up he needed. But, in fact, for this young man, the invitation was not wholehearted or sufficient. It invited him to enter the community, but not to participate in it as a true insider. And it did little to change the field so that others like him could join it.

Srivinasa Ramanujan Iyengar, better known simply as Ramanujan, was made famous by the movie about him, *The Man Who Knew Infinity*, based on a book of the same name by Robert Kanigel. Ramanujan lived and died almost one hundred years before Tao. Like Tao, he made his most significant contributions to the field of number theory. Ramanujan's approach to math was so novel that mathematicians are still trying to parse his notes to this day. And his background was so foreign to the Western mathematicians of his time that many found themselves giving him the same title as Tao: genius.

Ramanujan came from southern India to work as a mathematician at King's College, Cambridge, in 1914. He was invited there by the mathematician G. H. Hardy, one of the most esteemed mathematicians in the world at the time. Ramanujan spent five years in Cambridge until he fell ill and returned home to India, where he died in 1920.

That Ramanujan was able to get an invitation from Hardy to join the mathematical community at Cambridge is remarkable. Ramanujan was born in 1887 in Kumbakonam, a city in what is now the Indian state of Tamil Nadu. Kumbakonam's fantastically colorful Hindu temples could not be more different from the angular, dark stone cathedrals of England. Ramanujan's childhood was also a sampling of the traumas that many children and families faced in southern India at that time. His family was not well off. Of the six children his mother gave birth to, only three survived past infancy.

Ramanujan had no formal mathematical training. He studied math at the University of Madras, but his education had not prepared him to do mathematical research. Instead, his coursework prepared him to be an accounting clerk. He held a job as a clerk at the Port of Madras, now Chennai. Nonetheless, he conducted mathematical research, inventing his own system of notation. He spent much of his spare time working through advanced problems in pure mathematics. And he thought he was making some important discoveries.

Ramanujan wanted to share his research with the wider world. So, in 1913, he made a bold move: he mailed letters full of his amateur mathematical work to several prominent British mathematicians. When Ramanujan posted those letters in 1913, he had no idea how they would be received. And by all accounts, those who received his letters were shocked by the mathematical promise they showed.

Hardy then made an equally bold move: he invited Ramanujan to join him in Cambridge. Hardy's colleagues had their doubts about Ramanujan. To these well-educated British men, ensconced in their Cambridge towers far away from India, Ramanujan was an unlikely mathematician. They were in awe that someone from Ramanujan's background could produce such sophisticated mathematical work. But to them, he was almost too strange, and his work seemed too surprising to be taken seriously.

Kanigel's language in his biography of Ramanujan captures some of the wonder that the British mathematicians must have felt upon seeing an Indian man working among them. "Ramanujan was a man who grew up praying to stone deities," Kanigel writes, "who for most of his life took counsel from a family goddess, declaring it was she to whom his mathematical insights were owed." In the eyes of the Cambridge dons, Ramanujan was a strange foreigner who did not fit into their community. Ramanujan was a relatively poor man and unaware of many Western cultural practices that those mathematicians took for granted. For instance, he did not know that a person could sleep underneath blankets for warmth. After all, southern India is one of the hottest places on Earth. He had not experienced cold weather, so he was not familiar with the ways the English had of keeping warm. And yet, while in Cambridge, Ramanujan collaborated with two of the greatest British mathematicians of his time, Hardy and John Littlewood. Ramanujan published several mathematical papers during his short time in Cambridge. He was even made a fellow of the Royal Society.

To these British mathematicians, Ramanujan could have seemed like nothing less than a genius. It was the only way they could account for his mathematical knowledge, given that he had not been raised and educated in the ways that they had. The famous mathematician and philosopher Bertrand Russell commented that on the day after Hardy received Ramanujan's first letter, he "found Hardy and Littlewood in a state of wild excitement because they believe they have found a second Newton, a Hindu clerk in Madras making 20 pounds a year." It seemed like a miracle that could only be described with the magical term of genius.

He was called a second Newton. He published papers in collaboration with some of the world's greatest mathematicians, with no formal mathematical training. Based on these accomplishments, Ramanujan's story sounds like one of success. But was it?

Hardy may have invited Ramanujan to join him as a colleague at one of the top mathematical institutions in the Western world. But what was Ramanujan's role among the mathematicians at Cambridge? Was he truly welcomed to participate in the community of mathematicians? Or was he just invited to take a seat at a table where a conversation was already in progress, where he was not a full participant but a somewhat bewildered outsider? If we peel back some of the layers of Ramanujan's story, we see that it was not as rosy as it may at first seem.

Ramanujan pursued the mathematical research at Cambridge that he had started in India. He brought with him notebooks full of calculations and proofs, all written in mathematical language that he had developed himself. Ramanujan and Hardy spent much of their time going through those notebooks together. And the notebooks contained some real mathematical gems.

Much of Ramanujan's work was in number theory, the same field pursued by Tao. One of Ramanujan's most impressive achievements was a calculation of something called the "partition of two hundred." This is the number of ways that the number two hundred can be written as a sum of positive whole numbers.

You can probably think of many ways to write two hundred as a sum of positive whole numbers: 199 + 1, 198 + 2, 197 + 3, and so on. We could go on like this for a while. The ease with which you can generate a relatively large number of ways to make two hundred should give you a sense for how large a problem this is.

It turns out that there are nearly four trillion ways to write two hundred as a sum of positive whole numbers. Remarkably, Ramanujan made this computation entirely by hand. Mathematicians in Ramanujan's time didn't have calculators or computers. His computation was an amazing feat.

But no matter how amazing the calculation, the ability to do calculations does not make a mathematician. And here is where we begin to get hints that while Ramanujan's invitation to join the mathematical community at Cambridge may have gotten him to the table, he was not truly welcome to participate in the discussion. Some mathematicians have commented that Ramanujan's work feels like a collection of mathematical facts. Ramanujan was in fact skilled at computations. After all, he worked as an accounting clerk, basically a human calculator, before coming to Cambridge. Many of his findings, like the partition of two hundred, are computational. But mathematicians want theoretical proofs, not merely clever computations. The Cambridge mathematicians did not always understand Ramanujan's work to contain such proofs. So, although what they saw as Ramanujan's computational prowess impressed mathematicians, they were looking for something more.

As an essentially self-educated mathematician and an outsider to the mathematical community, Ramanujan did not know which of his discoveries might be new to the mathematical community. Much of the work in Ramanujan's abundant notebooks was actually not new. Ramanujan had been cut off from the mathematical advancements taking place in Western Europe until he arrived at Cambridge. He had no way of knowing what was modern mathematics and what was passé. It was remarkable that a self-educated man had been able to independently develop many of the proofs that mathematicians had discovered in the past. But Hardy and his British colleagues quickly identified as old news many proofs that Ramanujan had thought were original. Even worse, a lot of what Ramanujan had done in his notebooks was wrong. As Kanigel puts it in his biography, Ramanujan followed his intuition. But his intuition often led him astray. Hardy thought that only about one-third of what Ramanujan brought with him to England was mathematically valuable.

That one-third was enough to fuel several years of collaboration with Hardy. But even Hardy couldn't help lamenting what might have been, had Ramanujan received more formal mathematical training when he was younger. Hardy said, "The years between eighteen and twenty-five are the critical years in a mathematician's career, and the damage had been done. Ramanujan's genius never had again its chance of full development." Ramanujan clearly had potential to succeed in the world of mathematicians, Hardy and his colleagues seemed to say. They were impressed that he'd made it this far. But he could never fully live up to it. He had too many handicaps because he had not been conventionally educated in Western methods. Hardy called Ramanujan a "poor and solitary Hindu pitting his brains against the accumulated wisdom of Europe." Not a fair competition and not one that Ramanujan could win.

Ramanujan died with promise unfulfilled, in the eyes of Hardy and his Cambridge colleagues. But that is not the end of the story. Since Ramanujan's death, mathematicians have begun to find deeper significance in work once written off as too exotic. Some of his work that had been seen as behind the times was now starting to seem ahead of his times. The real significance of much of Ramanujan's work was not seen until mathematicians related it to more relevant contemporary problems. Ramanujan, it seems, had a knack for picking smaller, calculation-based problems that held the kernels of theoretical significance within them. Ramanujan's impressive partition calculation, for instance, improved on previously known methods for computing partitions. It led directly to the method that computers use to perform that calculation today. What looked to Hardy and others as computational wizardry incorporated significant theoretical developments.

Mathematicians today still do not fully understand all of Ramanujan's work. Bruce Berndt is an expert in Ramanujan's notebooks. He's spent years studying them. Even he has said, "I still don't understand it all. I may be able to prove it, but I don't know where it comes from and where it fits into the rest of mathematics." So, Ramanujan's work seemingly holds more substance and significance than Hardy and his British colleagues noticed at the time.

Hardy certainly valued Ramanujan's mathematical work. He respected Ramanujan enough to invite him to Cambridge and collaborate with him. But he never fully understood Ramanujan as a person or his work. How was it that Hardy was unable to see some of the mathematical riches in Ramanujan's work?

The answer to this question lies partly in British mathematicians' expectations of what Ramanujan would do when he arrived in England. Hardy and his colleagues pushed Ramanujan to learn the practices of Western mathematics. They did this for good reasons. Ramanujan wanted to publish in British mathematics journals and be appointed to higher positions in British mathematics institutions. Hardy wanted to help him achieve these goals. The idea was that if Ramanujan wanted to make it in the world of British math, Ramanujan had to learn how the British did math.

But Hardy and his colleagues may have been mistaken that the way for Ramanujan to contribute to British math was for him to adapt to conventional British ways—instead of British mathematicians learning some of Ramanujan's ways. They may have been overly biased against math that they found unfamiliar. This bias likely stemmed in part from a natural aversion to having to change their conventional ways of doing things. But it also likely arose from negative stereotypes of the intellectual capacity of people who weren't British. Those negative stereotypes fed a vicious, systemically ingrained cycle: because the British did not think that Indians could or should be mathematicians, they did not support educational institutions that taught formal mathematics in India. As a result, students like Ramanujan didn't learn formal Western math. So, when these students tried to show their mathematical talent to great British mathematicians, they were denigrated for not being familiar enough with that math. They were considered underdeveloped. Which fed back into the low expectations of Indian mathematicians.

In 1914, when Ramanujan set out for Cambridge, India had few institutions of higher learning. It had none that could support the kind of abstract mathematical research that Hardy did. This is not because India lacked a history of abstract mathematics. India had a long, rich history of mathematics before the British claimed it as a colony.

What we now call India was once a collection of kingdoms that rose and fell over thousands of years. Scholars flourished in many of those kingdoms, including under the Cholas, who founded Ramanujan's hometown of Kumbakonam. But the Europeans who colonized India for their own gain did not consider the development of scholars in India as beneficial to their rule. They did not value or support traditional Indian scholarship and did not promote high-level Western scholarship in India.

The University of Madras, the preeminent university of southern India in Ramanujan's time, could not support Ramanujan and others like him to become professional mathematicians. This was one of the many ways that the British subjugated their Indian subjects during colonial rule. The infrastructure that the European powers built in India was not for the benefit of Indians. Putting it bluntly, it was meant to help European colonizers make money.

To make money, the colonizers had to train Indians to work for them. Toward this end, they built universities. But they limited what and who could be taught at these universities. They took advantage of India's existing caste system to make decisions about who would get to attend those universities. Ramanujan was able to study math at the University of Madras because he was a Brahmin, the highest caste in Indian society. Where Ramanujan was from, a person's level of education and job prospects were entirely determined by gender and caste. Although his family was not well off, Ramanujan could attend school, learn English, and dabble in math because he was a Brahmin. But while British universities such as Cambridge were meant to nurture young British men to become scholars and academics, the universities that the British built in India had a much more modest mandate. Indian universities trained Indians to be government officials, managers, and accountants. The British did not need Indians to become scholars. They needed Indians to become better colonial subjects and workers.

Ramanujan had the great privilege among Indians to study math at the University of Madras. In receiving this opportunity, Ramanujan was lucky. Consider the life of his wife, Janaki, for comparison. Janaki was married to Ramanujan when she was just ten years old. When he died in 1920, she was only twenty-one. Despite living until the age of ninety-five, she followed custom and never remarried. While seeing an Indian boy become a famous British mathematician might have required a stretch of the imagination in the early twentieth century, seeing the same path for an Indian girl requires imaginary leaps and bounds.

The math that Ramanujan learned at the University of Madras was not meant to help him develop into a mathematician. In order to have his work noticed by the community of people who could help him, Ramanujan had to take the extraordinary step of sending unsolicited letters to prominent British mathematicians, men to whom he had never been introduced. Except for Hardy, these mathematicians rejected his advances out of hand, probably at least in part because he was Indian. But Hardy's support was not enough to bring full recognition to Ramanujan's work. Ramanujan's education did not prepare him to communicate with British mathematicians. And British mathematicians were not open-minded enough to really dig into the work that Ramanujan sent them.

The British mathematical community did not fully accept Ramanujan as a mathematician. But this is not the only way that Hardy's invitation did not fully invite Ramanujan to participate in the community of mathematicians. The British mathematical community also did not fully accept Ramanujan as a person.

Ramanujan was one of a small number of Indian students in Cambridge during the early twentieth century. The cultural practices that Cambridge students and professors took for granted were foreign to Ramanujan. Even worse, they conflicted with his own cultural practices and religion. For example, Ramanujan was a vegetarian. But all of the food served in the communal dining hall, where Ramanujan's colleagues had most of their meals, had meat in it. Ramanujan was cut off from this central opportunity to bond with his colleagues—and to eat dinner. So, he prepared his own meals. But when World War I broke out soon after his arrival, he struggled to find the ingredients he needed because of heavy food rationing.

Ramanujan was starved—literally for food, but also figuratively for someone who truly understood his mathematical thinking and for human connection. Even Hardy, who interacted with Ramanujan regularly, did not develop more than a professional relationship with the young man he had taken under his wing. Hardy claimed that they could not be friends because they were too different. He said, "Ramanujan was an Indian, and I suppose that it is always a little difficult for an Englishman and an Indian to understand one another properly." Hardy could and should have done better.

The hardships of being an outsider took their toll on Ramanujan. Kanigel writes that while on the outside Ramanujan appeared to be working enthusiastically on his mathematics, "inside, Ramanujan was like a checking account from which funds are only withdrawn, never deposited. Doing mathematics took vast personal energy. So did adjusting to his new life in England, as anyone will attest who has ever tried to penetrate a foreign culture. Together they drained his physical and emotional reserves. Eventually, the account must run dry." After a few years in England, Ramanujan fell sick with tuberculosis. In 1920, he died.

What Ramanujan did to become a mathematician was extraordinary. He took great risks in sending his work to strangers on the other side of the world, and then in going to live with them. He suffered mathematically, socially, and physically. That a person from Ramanujan's background should have to struggle so hard to get a chance, whereas a person from Tao's background should have his way paved for him, is unfair and unproductive. While Ramanujan made some impact in his time and continues to do so today, how many other promising mathematicians have been lost because of the hardships they could not overcome? Although Ramanujan made it in some respects despite the hardships, we should not wish what happened to him on another person. And while Ramanujan's story is inspiring, it inspires in part because it is rare. The world may be lucky enough to produce two or three

Ramanujans, but certainly not one hundred or one thousand, as it could if students were properly taught and encouraged.

It is unreasonable to expect other aspiring mathematicians who are viewed as outsiders by the tight-knit mathematical community to replicate the extraordinary steps Ramanujan took to obtain recognition for his work. Such an expectation only perpetuates the problem. So does the expectation that someone like Ramanujan could thrive in a world he was invited to visit but was structurally prevented from fully participating in.

Math and Democracy

Ramanujan's story shows that if we want to diversify the community of mathematicians, it is not enough to invite promising mathematical minds from underrepresented groups to join the field. An invitation alone is not sufficient to enable newcomers to the participate fully in the work of the community. If the community does not change in response to the new ideas and ways of interacting that newcomers bring with them, those newcomers will remain outsiders in all but name. And they will face hardship as a result. The hurdles they have to jump in order to become fully participating members of the community are much higher than those faced by other students. Sometimes those hurdles are impossibly high.

That's why BEAM does more than simply help students from backgrounds that are underrepresented in math to get into existing math enrichment programs and elite New York City magnet schools. Students who graduate from the BEAM summer program and workshops receive a safe space in which to develop their identities as mathematicians. BEAM classes help students learn to use their voices to push the culture of mathematics to accept them and their creative ideas. This is just as important as the opportunities BEAM provides to help those students get into that culture in the first place, and in many ways much harder to do. The complexity of it is best captured by the stories told by BEAM teachers.

Three BEAM staff members sit around a small table tucked in a noisy, shared midtown Manhattan office late in July, reminiscing about summers past. They are a summer teacher, Ben Blum-Smith; the program coordinator, Ayinde Alleyne; and Zaharopol's second-in-command and director of programs and development, Lynn Cartwright-Punnett. They tell stories about mathematical moments starring their students. Their stories are filled with drama that you only hear from the most passionate of teachers.

At this meeting and on his blog about math and teaching, *Research in Practice*, Blum-Smith, a PhD student at the Courant Institute of the Mathematical Sciences at New York University, tells a story from the summer of 2013. The story is set in his number theory class. Going into the summer of 2013, Blum-Smith had been trying to come up with ways to encourage his students to get into the practice of summarizing each other's thoughts. Not evaluating those thoughts, but restating them. Mulling them over aloud for everyone else in the class to hear. Deciding for themselves which parts were most worthy of emphasis or further inquiry.

There are many reasons why math teachers might want their students to summarize each other's thoughts. Some of these reasons have more to do with controlling behavior than anything deeply mathematical. For instance, if kids can summarize each other's ideas, it shows that they have been paying attention.

But this was not what Blum-Smith was getting at. Instead, he was stewing over an idea for math teaching that he now calls the "democratic process." He was wondering how to set up a math classroom in which the students would work together to formulate deep mathematical ideas that contained all of the thinking unique to each of them. This is important for all math students. But he thought it was particularly important for BEAM students because they needed to consider themselves as powerful mathematicians.

As people who had something valuable to contribute in a mathematical community.

How could he encourage these students to state their mathematical ideas in ways that truly expressed them? And how could he then get other students to explore those ideas? To decide which aspects of those ideas spoke to them and which were troubling? And then work together to make them better?

Blum-Smith wanted to set up a classroom that would encourage his students to express their mathematical ideas in ways that were responsive to the voices of everyone in the room. These students were accustomed to mathematics being static and unresponsive. But Blum-Smith knew that this did not have to be true. To convince these students to express their brilliant mathematical ideas, he knew he would have to provide them with experiences they perceived as powerful for themselves. One way of accomplishing this goal, Blum-Smith realized, was by having kids routinely restate each other's thinking.

The story Blum-Smith told was about the day he fully realized the power of his teaching idea. On that day, he and his students were working on an important problem in number theory, proving that there are infinitely many primes. Euclid long ago proved this problem, but it remains central to number theory today, and figuring out its proof is a great mathematical exercise for students.

That there are infinitely many primes may sound obvious. Numbers go on forever. So why should the primes have a limit? But the infinitude of primes need not accompany the infinitude of numbers in general. If all you knew was that numbers in general go on forever, you could imagine a world in which all numbers are made out of some finite set of primes.

All numbers—prime and not prime—can be broken into prime factors. Fifty-six is not prime. If you divide fifty-six into the set of prime numbers that total fifty-six when multiplied together, you get $7 \times 2 \times 2 \times 2$. Dividing fifty-six up in this way is called finding its prime factorization. Seven is prime. Its prime factorization is just seven. Each number has its own unique prime factorization. This is another seemingly obvious statement about numbers proven by Euclid. But it's so important to number theory that it is called the Fundamental Theorem of Arithmetic. But what if after some point all the numbers could be factored into the same set of primes, just multiplied together in different ways?

If you were in the business of finding new, larger primes (as a number of people are), you might worry that this was the case. Sometimes big numbers have small prime factors. And sometimes big numbers use the same prime factors many times. You don't need a diverse set of prime numbers to make lots of big numbers. For example, the prime factorization of 7,168 is: 7,168 = 7 × 2 × 2 × 2 × 2 × 2 × 2 × 2 × 2 × 2 × 2.

Furthermore, as numbers get bigger, the primes thin out. There can be large gaps between primes. For instance, mathematician Harvey Dubner, who is one of those mathematicians in the business of looking for primes, found a gap of 12,540 between the following enormous prime and the next enormous prime:

> 102,811,585,161,859,662,929,133,834,596,957,332,561,175,
> 592,034,953,605,055,721,223,249,969,500,653,795,121,975,
> 853,179,617,590,006,903,289,133,192,447,178,976,880,198,
> 220,637,378,125,686,339,726,137,874,956,095,491,930,654,
> 497,693,978,715,833,794,999,935,477,468,391,789,508,344,
> 449,541,406,347,900,355,427,290,700,854,945,945,853,
> 825,193,979,651,314,099,863,832,554,824,576,338,414,
> 272,502,493,678,448,947,860,165,143,562,942,794,028,
> 966,359,380,108,925,040,409,462,881,632,270,278,716,570,
> 882,306,451,587,569

A gap of merely 12,540 between this monstrosity and the next may feel insignificant. If the next prime were written here, you would probably get bored reading it before you noticed that it was a different number. But the process of finding primes makes this gap much more significant. To find primes, mathematicians try dividing numbers by known primes. If a known prime divides a number, the number isn't prime. Think of how many primes you would have to try dividing into each of those 12,540 numbers before you knew that they weren't prime. After ten thousand non-primes you might begin to worry: What if I have found the last one? So, it's important to mathematicians to know for certain that primes go on forever. The proof of this fact is fundamental to number theory.

Blum-Smith's BEAM students were going to construct Euclid's proof of the infinitude of primes. The students developed this challenge for themselves, just as Blum-Smith hoped that they would. Like Euclid, they arrived at this question by chasing down the path forged by mathematical curiosity and good problems. And by the end of one week, they'd solved the problem. Euclid's ingenious proof for the infinitude of primes lay firmly in their grasp. But their success in finding the proof is the beginning of Blum-Smith's story, not the end. His story is about democracy in math, not just about finding right answers.

For at least five minutes, Blum-Smith entertains his audience of fellow BEAMers with the thrilling tale of the formation of the proof. Four days into the class, Blum-Smith says, a student named Jaden spat the proof out. It was fully formed, but in a series of convoluted sentences. Blum-Smith understood what Jaden had done. But he had a hunch that most of Jaden's fellow students did not.

"People were not with him," Blum-Smith says. "They were *primed* for it"—this math joke elicits a "buh-dum-chhhh!" from Cartwright-Punnett and the most quiet of ha-ha's from Alleyne. But they were not quite there yet.

It isn't too hard to get one student in a room full of mathematically bright kids to prove the infinitude of primes. But it's much harder to get all of the students to understand the proof. It's even harder to get all of the students to feel that they, as a community, developed the proof. Blum-Smith's goal was to create a community of mathematicians. Toward this end, Blum-Smith orchestrated a far more complex process for solving math problems than just having one student come up with the answer while the other students applauded. This was Blum-Smith's democratic process.

If Blum-Smith had simply told Jaden and his classmates that Jaden's proof was correct, the process would not have been democratic. Jaden and Blum-Smith would have participated in it. But everyone else would have been a passive bystander. Other students might have had ideas that were relevant to the proof. Just because those other students hadn't verbalized a full proof yet didn't mean that they didn't have something valuable to contribute.

Blum-Smith's democratic process started with summarizing. Tuliya, one of Jaden's classmates, summarized Jaden's argument "much more clearly than he said it." ("Of course she does," Cartwright-Punnett chimes in.) But she left out a key part of the proof—which Jaden, while listening to her, did not catch.

According to Blum-Smith, Tuliya began by asking, "What if the primes do end—and what if I have a list of all the primes that exist?" This question may sound counterproductive, since it seems to be calling for an answer that is the opposite of what they are trying to prove. Tuliya and Jaden want to show that there will always be another prime. So why would Tuliya ask us to imagine that she can write all of the primes on a finite list?

Tuliya and Jaden's strategy of assuming the opposite of what you really want to prove is a common practice in mathematics. If you want the primes to go on infinitely but cannot think of a way to show that they do, you can alternately play out the scenario that they do end. If your initial hypothesis that they do not end is correct, you will inevitably arrive at something clearly false and absurd. And if the assumption that there are finite primes leads to clearly false and absurd mathematical conclusions, then the primes cannot end. They must go on infinitely. This kind of back-alley proof is called a proof by contradiction. A proof by contradiction basically says, "The thing I want to be true cannot be false, so it must be true."

Once Tuliya and Jaden had imagined that all of the primes could be written on a single list, they performed a clever trick. They took all of the numbers on the list, multiplied them together, and added one. They now had a new number. This number was much too large to be on the initial list. And it had a distinct possibility of being prime itself.

An example shows how this works: say you thought the only primes that existed were two, three, five, seven, eleven, and thirteen. Following Tuliya and Jaden's method, multiply those numbers together:

$$2 \times 3 \times 5 \times 7 \times 11 \times 13 = 30{,}030.$$

And add one:

$$30{,}030 + 1 = 30{,}031.$$

The result, 30,031, certainly is not on the initial list of primes. Remember, only two, three, five, seven, eleven, and thirteen are on the list. Even more importantly, none of the initial primes divide into 30,031. A number that is only one larger than a multiple of two, three, five, seven, eleven, or thirteen can never be a multiple of any of those numbers. The nearest multiple of two is still one away. The nearest multiple of three still two away. And so on. If none of the primes divide into 30,031, then could it be . . . prime itself?

Tuliya said, "YES! Any number made by multiplying a list of primes and adding one must be prime!" And she wrapped up her proof. For, if she had created a new prime, then she had shown the absurdity of the existence of a finite list of prime numbers. She had also developed a way to make a new prime number from a set of other prime numbers. She had, in a sense, a recipe for primes. It was an elegant piece of mathematical reasoning. But was it right?

When Tuliya had finished her summary, Blum-Smith, in true democratic style, took a class vote. Blum-Smith asked the class, "Raise your hand if you feel that you understand the idea that Jaden put forth that Tuliya is summarizing." Two-thirds of the kids raised their hands.

Then he said, "Leave your hand up if you also find the idea convincing and you now believe the primes don't end." Blum-Smith and a few kids put their hands down. In summarizing Jaden's argument, Tuliya had missed an essential step. Blum-Smith wasn't sure if the other kids in the class, Jaden included, had noticed.

Blum-Smith could have pointed out the mistake himself. Maybe another teacher would have. But that wouldn't have been democratic. Instead, he left space for the group to participate. Jaden had produced the whole proof, but not everyone understood it. Tuliya had produced a proof that most students understood, but it missed a key piece. It would take the whole team coming together to produce a complete proof that everyone could get behind.

What was the missing piece? Perhaps you've already noticed it: 30,031 is not prime. Its prime factorization is 509 × 59. Tuliya and Jaden's method easily could have produced a prime number. If their list had only included two, three, five, seven, and eleven—without thirteen—they would have made the number 2,311. Which is prime. Multiplying many primes together and adding one is a method that mathematicians frequently use when attempting to make prime numbers. As in the case of 30,031, however, it does not always work.

But all is not lost. True, 30,031 may not be prime. But in its prime factorization we find two new prime numbers, neither of which were on the initial list. Tuliya's logic that none of the initial primes could divide her new number still holds, even if the number is not prime. Her recipe for primes works. Only sometimes it produces one new prime, and sometimes two.

Blum-Smith's students closely examined Tuliya's summary. Eventually, they constructed a proof that stood out for being not only accurate but also reflective of the ideas of the community that produced it. Blum-Smith loves this story precisely because it highlights the key roles that the democratic process and community play in mathematical discovery. "The kid who had the whole thing in the first place," he says, referring to Jaden, "that didn't settle it. What settled it was this whole process." A process of deconstruction and reconstruction, bringing voices together to determine what the whole community valued in a proof. The proof produced by the group working together is a proof that shines. But only because everyone's voice was included it its production.

How does a group of mathematicians reach agreement on a proof? With Blum-Smith's story in hand, the process seems much less straightforward than Poincaré describes. This is true for the discoveries of professional mathematicians as well as middle school students. It at least requires more than an innate sense possessed by a single mathematician. It requires a community of mathematicians who collaboratively built a culture of sharing, listening, and probing each other's ideas. Each idea benefits from having been stated in one voice and restated by another. The more voices, the better.

Even for Blum-Smith, a professional mathematician, identifying why components of a mathematical discovery are important requires empathy as well as a mathematical sense. "I always struggle," Blum-Smith shares toward the end of his story. "As a mathematician reading other people's proofs. It's always, like, trying to keep in mind, why are they arguing this sentence?" To try to understand why another mathematician values a particular finding, Blum-Smith puts himself in conversation with the mathematician who wrote the proof, much as he put Jaden and Tuliya in conversation with their classmates. The conversation he has is as much about individual mathematical facts as it is about a cultural sense that mathematicians have about what matters in their field.

This deeply cultural conversation among mathematicians is what makes mathematics significant. Examining the role of this conversation in doing math helps us understand the struggle faced by New York City teenagers who belong to underrepresented groups, prodigies in colonized territories such as Ramanujan, and others trying to join the mathematical community. Their struggle comes not from a lack of talent, but from a lack of history. Those who belong to that culture learn how to engage with it over a long period of time, often starting as children when they join their first math team or talk about math on a walk with a parent. Given how hard it is to learn to be a mathematician, it's easy to see how the world of mathematicians could be so homogenous and insular.

But homogeneity among mathematicians does not benefit mathematics. Jaden's proof was improved with the participation of his classmates' diverse voices. In the same way, the field of mathematics itself improves as new voices pose and solve problems. A democratic process gives people more power in making new math knowledge and in making math that empowers more people.

A KNOWLEDGE OF MATH also gives more people more power in the democratic process. At least, that's what civil rights advocate and math educator Robert Moses thinks.

The story of BEAM shows us one way that mathematicians can help diversify their field. They invite people who have traditionally been excluded from the field to join. But they also give them space and support so that they can truly participate in making math.

But making the world of mathematicians more diverse isn't the only challenge that math educators face. Not everyone who studies math will become a mathematician, but everyone has to study math. Knowledge of math is essential to getting a job in our modern technology-driven economy. Even if students do not plan to go into careers involving math, it is impossible to graduate from high school and college without passing math classes, which can be a barrier to students trying to get a good education. And without a good education, it's difficult to have power in our society.

Here's how the math barrier works: many colleges require students to take a math class even if they do not intend to major in a STEM field. But students often arrive at those colleges unprepared for even the lowest college-level math classes offered because of inferior high school math education. As a result, they fail the college's math placement test and are not allowed to take a college-level math course. They must first pass remedial courses. Students have to pay for these courses, just as they would any other college class. But remedial courses often do not count toward a degree. Remedial students are stuck in limbo, paying for college but not progressing through it.

The strain that a student is under once she arrives at college and learns that she has to pass remedial courses before she can even begin taking classes for credit is immense. Most students expect their high school math classes to have prepared them for college. At the very least, high school should prepare students to quickly pass remedial courses. But a study of Texas college students enrolled in remedial math courses found that only thirty-three percent of the students passed those classes. And only eighteen percent of those students went on to complete their first college-level course. Remedial math courses drain the bank accounts of college students and often do not bring students any closer to graduation.

This situation affects black students at a higher rate than it affects students of other races and ethnicities. Inadequate math education is a problem in many schools in the United States, but it is especially a problem for black students. Although legal segregation is a thing of the past, it is still the case that black students overwhelmingly attend predominantly black schools and that these schools are less effective than most white schools. The results are evident in the problems that black college students have with math. Fifty-six percent of black first-year college students in the United States were enrolled in remedial courses in 2016. The rate of remediation for all students was nearly fifteen percentage points lower. While substandard math education in high school is a problem for many students, these statistics show that it is a bigger problem for black students. Black students are effectively paying more for their college educations and getting less. They are being left behind in math at a higher rate than their white peers. This disadvantage decreases their access to good-paying jobs in technological fields, to careers in advanced STEM fields, and to the important problem posing and solving that knowledge of math affords.

According to Robert Moses, this is a civil rights issue. Moses once led efforts at the Student Nonviolent Coordinating Committee (SNCC) to register black Mississippians to vote in the 1960s. Today, he is still working to enfranchise black Americans. Moses now works in a new space, math education, one that might seem far from voter registration. But not for Moses. He argues that the high rate of math illiteracy among black Americans, a result of many years of educational inequity, is now the biggest hurdle they face in achieving full enfranchisement in the United States. Math illiteracy, according to Moses, does as much to prevent blacks from participating in American democracy today as voter suppression did in the 1960s. And he's trying to fix that.

Like the mathematicians at BEAM, Moses realized that black students need a place to learn math where they feel affirmed, safe, and valued. He also realized that he needed to work with the whole community, not just individual students. He needed parents and teachers to join him in the fight for high-quality math education for their students. So, Moses did what he does best: he organized the community. His math-centered community-organizing project is called the Algebra Project. Begun first in Cambridge, Massachusetts, the project now focuses on predominantly black communities in cities and in the southern United States.

For many civil rights advocates, making progress is a process of pushing to enlarge the space in which black people can thrive. But it's difficult to put pressure on a space if it is completely closed off. What you need to get started, Moses writes in his autobiography, is a "crawl space." You need a small territory tucked away in the nether reaches of the space you're trying to broaden. There, you can safely begin organizing those who will work with you.

Moses chose the crawl space of his own daughter's middle school math class. Moses had been doing math with his daughter Maisha at home for years. Now that she was entering eighth grade, Moses found that the math curriculum covered many of the things she already knew. He thought she was ready for algebra. But there was no way for her to learn it within the math classes available at her school.

So, Moses made an offer to Maisha's teacher: he would teach algebra to Maisha and any other student in her class who wanted it. Maisha's teacher agreed. And Moses had his crawl space.

Why was middle school algebra such a good crawl space for Moses? Moses had identified a trend as he looked at middle schools and high schools around the country. Students who attended elite private and public schools got to take algebra at younger ages. Taking algebra earlier prepared them to go further in school, and get there faster. It prepared them to learn more about science and technology, the linchpins of the modern economy. Algebra prepared them, in Moses's words, "to become America's leaders." But, says Moses, "all other schools belonged to an era in which work and preparation for work were defined by factories and assembly lines." The students of color who often attended these less-than-elite schools were not being prepared to join in the modern economy. And lack of economic power for marginalized groups results in lower status and less power in society. Less say in the decisions that affect them every day. Being cut out from decision-making because of a lack of education. This was oppression. Moses argued that he could trace that oppression all the way back to the decision not to offer algebra classes at middle schools that predominantly served students of color.

So, Moses taught algebra to his daughter and her friends. Moses's algebra students went home to their parents raving about their new math class. And the parents took notice.

When asked by a reporter, "How do you organize?," Moses replied, "By bouncing a ball. You stand on a street and bounce a ball. Soon all the children come around. You keep bouncing the ball. Before long it runs under someone's porch and then you meet the adults." That's precisely what happened in Maisha's school. The parents began to wonder, What if *our* eighth graders could also learn algebra? What opportunities would be open to them? More broadly, they wondered, What if the parents and children had a say in what math was taught at school?

These ideas were revolutionary for the parents of students of color Moses encountered. In his autobiography, Moses quotes a letter that one parent, Shirley Kimbrough, wrote to the other seventh-grade parents at their school. After attending a meeting about the academic challenges faced by "minority students" at the school, she wrote, "Generally they were not doing as well as non-minority kids in math and science. Why?" She noticed that white students often got to take algebra at younger ages than students of color. Learning algebra helped them get ahead. Seeing what Moses's algebra program was doing for Maisha and her friends encouraged her to ask a revolutionary question: Why shouldn't all students of color also have this opportunity?

Parents like Shirley began to organize. Their action led Moses's school to open his algebra program to all students, not just to high achievers. Moses began offering Saturday algebra classes to parents who wanted to help their kids with their algebra homework but had never learned algebra themselves. Steadily, Moses's algebra initiative grew. From one crawl space in his daughter's eighth-grade math class, Moses's program expanded to schools with large populations of black students in cities such as Chicago, Oakland, and Indianapolis, and in small towns in the rural South. Everywhere he went, Moses organized communities in similar ways: first do good by the kids, then get the attention of the parents, and finally transform how schools teach math to its black students.

Moses writes that he "did not know that my concern for Maisha's math education would lead to the Algebra Project's raising questions about ability grouping, effective teaching for children of color, experiential learning, and community participation in educational decision making." He also did not know that the statement that *everyone* should learn algebra in middle school, including students of color who often ended up in a school's lowest tracks, would take hold across the country. It was a radical statement. But it needed to be said.

At the end of his autobiography, Moses quotes Ella Baker, one of the first organizing members of SNCC. Her mathematical play on the word "radical" is not the only thing that makes her statement apt for this story:

> In order for us as poor and oppressed people to become a part of a society that is meaningful, the system under which we now exist has to be radically changed. This means that we are going to have to learn to think in radical terms. I use the term *radical* in its original meaning—getting down to and understanding the root cause. It means facing a system that does not lend itself to your needs and devising means by which you change that system.

Baker's words encourage us to imagine a radical future for math and math education. The culture of mathematics does not always lend itself to the needs of all members of our national and global community. Math is used to pose and solve our most pressing social problems. But the people that the community of mathematicians leaves out are often precisely those who suffer at the hands of the most powerful social problem solvers. BEAM and Moses's Algebra Project are just two possible approaches to the problem of inequity in math education. What others can we create?

5 What Is Genuine Beauty?

Math and the Problem of PERCEPTION

Day and Night

The sculptures *Day* and *Night* exquisitely capture their namesakes. *Day* radiates in bright yellow, with sunray-like points. Its glow fills the room and entices visitors to look only at it, despite the multitude of other works of art vying for attention. *Night* retreats in dark blue, with craters that resemble those on the moon. You could miss it, sitting as it does, beside *Day*. But it, too, has gravity that draws you once you have noticed it. Together, they are a study in opposites: light and dark, sharp and smooth, extroversion and introversion. One captures the essence of the sun, and the other the soul of the moon.

Knowing nothing more about these sculptures, you might expect to find them in an art museum. Maybe a sculpture park. Certainly not at a math conference. But that is where I saw them. Their creator, Eve Torrence, is a full-time math professor. But she is also a pioneer in an unlikely genre: mathematical art.

What could math and art possibly have to do with each other? In many ways, math and art couldn't be more different. We often think of math as efficient, accurate, and useful. It should help us solve problems. Art has a different purpose. Art is about self-expression, probing the depths of the human experience and seeking higher purpose through beauty.

But many mathematicians have argued that the beauty of the work they produce is a crucial element in determining its value. In fact, mathematicians and artists are often on the same quest, searching for the meaning of beauty and trying to embody it in their work. Mathematicians and artists alike have lived and died in this quest for beauty. We are familiar with artists, writers, musicians, actors, and many others who devote their lives to the creation of artifacts that they hope others will deem beautiful. So do mathematicians.

How can something be both math and art at the same time? Let's take a closer look at Torrence's sculptures to find out.

I FIRST SAW *DAY* AND *NIGHT* at the Bridges Conference, a math conference that would strike the uninitiated as bizarre. If you strolled around without knowing who the attendees were, you might not guess that they were mathematicians. That's because of all the art.

Signs pointed to galleries full of paintings, sculptures, and mobiles. Downstairs, people learned to fold origami. Advertisements highlighted "Music Night," a performance of a play, and a mime show. Lectures about Islamic tile art and modernist sculptures filled the schedule. Sure, the lecturer for the modernist sculpture talk was the renowned University of California, Berkeley, mathematician Carlo Sequin. And his talk was peppered with words like "topological," "genus," and "order," unfamiliar words in an artistic context. And the songs sung at "Music Night" all had mathematical themes. Nonetheless, this still did not look like a math conference.

When you visit the Bridges Conference, you stand on the bridge between math and art. And what wonderful things you find there, including Torrence's sculptures. These objects are beautiful to look at, much like any piece of art. *Day* and *Night* drew me in from across the room in part because of their striking depiction of the cosmos.

But *Day* and *Night* also enticed me for mathematical reasons. In fact, *Day* won the award for best in show, an indication that the conference organizers thought it was both artistically *and* mathematically beautiful. "Mathematical art must include both mathematical ideas and artistic content," Torrence says. The mathematical artist "must use mathematics in the piece's design as well as make aesthetic choices in the construction of a piece." Torrence's sculptures did this exceptionally well.

So, how are *Day* and *Night* mathematically beautiful? They draw on mathematical contrasts. And the mathematical contrasts they draw on are similar to the contrasts that make them artistically beautiful.

Torrence made *Day* and *Night* by playing with the combination of two shapes that seem incompatible from a mathematical perspective but that she was able to make work together. Much as light and dark, sharp and smooth make for artistic contrasts, Torrence uses two shapes—hyperbolic paraboloids and Platonic solids—to draw mathematical contrasts.

A Platonic solid is an orderly three-dimensional shape. Each of its faces is perfectly flat. There are five different types of Platonic solids. Each is made out of regular shapes—such as a triangle, square, or pentagon—attached in the same way around each corner. *Day* and *Night* are dodecahedra, a Platonic solid made by gluing together twelve perfect pentagons.

A hyperbolic paraboloid, on the other hand, is an unruly curved surface. Looking from afar, you notice that it slopes upward heading north and south, and downward heading east and west—kind of like a saddle. But if you were an ant walking on a hyperbolic paraboloid, it would feel as flat as a piece of paper or the surface of the Earth. Hyperbolic paraboloids go on and on forever. Their curvature and infinite expanse make them hard to handle.

Each of the pentagonal faces of *Day* and *Night* is made of paper hyperbolic paraboloids. How do you make a curved surface out of a flat sheet of paper? And how do you attach such wavy surfaces together to make a solid with flat faces? These are the kinds of mathematical contrasts that *Day* and *Night* beautifully illustrate.

Torrence's paper hyperbolic paraboloids are actually approximations of the shape of a hyperbolic paraboloid. They aren't curved in the way a hyperbolic paraboloid is, but rather rely on the illusion of curvature that you get when you make many accordion folds in a sheet of paper. Approximating something curved out of something straight is an important but also baffling mathematical concept. The way that Torrence's sculptures illustrate it is mathematically creative and beautiful.

Fitting together the hyperbolic paraboloid approximations to make a dodecahedron took mathematical finesse. But this time, it took finesse of precision rather than approximation. The first time she tried to make the shape, Torrence says, "it failed miserably. It took me about six months to figure out the geometry to make this work." Torrence's first challenge was that hyperbolic paraboloids are nothing like pentagons. A dodecahedron is made out of twelve flat pentagons, but hyperbolic paraboloids are curvy and do not have five sides. Fortunately, Torrence had a strategy for dealing with that: arrange five hyperbolic paraboloids around a single point, making a shape like a flower with five petals. You get a structure resembling a pentagon in all the mathematically important ways. Even though it is covered in peaks and valleys, five hyperbolic paraboloids arranged in this way have five "corners" and five "edges." They also spin all the way around in five one-fifth-turn rotations. To a mathematician, if it spins like a pentagon, it is a pentagon.

Torrence's main mathematical challenge came from the shape of the hyperbolic paraboloids themselves. She only knew how to make hyperbolic paraboloids out of square pieces of paper. But while square hyperbolic paraboloids could make pentagons, they couldn't fit together to make a dodecahedron. The technical mathematical diagnosis for Torrence's square hyperbolic paraboloid problem is "hyperbolic paraboloids folded from squares cannot make a stable symmetric dodecahedron because they do not form rhombi with the correct ratio of the underlying rhombic triacontrahedron." In layman's terms, the structure they made was weirdly lumpy. So, Torrence had to do some complicated geometry. What shape would work? Torrence eventually found a rhombus that made the perfect hyperbolic paraboloid that in turn made the perfect dodecahedron. Precisely executed calculations are also mathematically beautiful. *Day* and *Night* capture this exceptionally well, too.

Mathematically as well as artistically, *Day* and *Night* are studies in contrasts. Curved and flat. Infinite and finite. Approximate and precise. Torrence's sculptures speak to the viewer in mathematical as well as artistic ways.

"Beauty Is the First Test"

A real piece of mathematical art, mathematical artist Eve Torrence says, "should give the viewer a deeper appreciation of both the art and the mathematics. The best mathematical pieces demonstrate deep mathematical ideas and give us insight into those ideas. They are both mathematically and artistically beautiful. You get double the joy."

But math need not be incorporated into a work of art in order to be considered beautiful. In fact, many mathematicians are drawn to the field because they think math is intrinsically beautiful, not because they think it is useful. For them, math and art are one and the same. These mathematicians do math to create beauty.

Many of the mathematicians profiled so far in this book have not held this view. They certainly enjoy math, and maybe they find it beautiful. But their work falls into the field of *applied mathematics*—the use of math in fields other than mathematics, such as anthropology, economics, politics, the law, and the sciences. They care about math because it helps them solve non-math problems. But there is another field of math called *pure mathematics.* While applied mathematicians often borrow ideas from pure math, most pure mathematics is not used outside of math.

In fact, pure math is usually developed with no applications in mind. And what many mathematicians who work in pure math love about their field is that it is application-free. They might even turn up their noses at their applied-math peers. Early twentieth-century mathematician—and Ramanujan's mentor—G. H. Hardy once wrote that "little of mathematics is useful practically, and that little is comparatively dull." According to mathematicians like Hardy, the real world is full of limitations. It's boring. Applied mathematicians are trapped there. But pure mathematicians revel in imaginary spaces full of exciting things like the infinitely expanding hyperbolic paraboloids that Torrence uses in her sculpture and pure mathematicians study as things in themselves.

Pure math isn't supposed to be useful. But that doesn't mean it doesn't address essential problems. Pure math takes up the same problems as art. It tries to answer one of the deepest human questions: What is beautiful?

Hardy was one of the most famous mathematicians to articulate this view of math as a quest for beauty as opposed to utility. His description of a mathematician's purpose in life has become a standard in the field. "A mathematician," he wrote, "like a painter or a poet, is a maker of patterns." And those patterns offer satisfaction in themselves, without any other purpose. In his widely read book *A Mathematician's Apology*, he offered a sarcastic pseudo-apology for the uselessness of most math. If pure math was useless, Hardy had wasted his time devoting his life to it. At least, that was the opinion of the pragmatic world of practical utility. But, he argued, the beauty created by the art of pure mathematics makes up for its uselessness. If you think painting and poetry are good uses of people's time, then pure math is, too.

Hardy is the most well known and widely cited modern mathematician to compare pure math to art. But he was by no means the first. The ancient Greeks connected math and aesthetics. For example, the Pythagoreans, as the ancient Greek followers of Pythagoras called themselves, based their whole philosophy of life around mathematical beauty. Numbers and patterns organized the universe, they thought. And not in a practical way, as one might think about math organizing science. But in an order-bringing, aesthetically profound way. In their view, math ruled the universe, and the secrets of the universe could be found in math.

Hardy was not the last mathematician to talk about mathematical work as art. Most of the modern mathematicians interviewed by Marianna Cook in her book *Mathematicians: An Outer View of the Inner World* cite beauty, art, imagination, and elegance in describing why they love math. The practical results developed from their work play second fiddle to the beauty of the work itself in its unapplied form.

David Mumford, who won the prestigious Fields Medal for mathematical achievement, describes pure mathematics as "secret gardens, places where I could try to grow exotic and beautiful theories." According to algebraic geometer János Kollár, pure math is "the most romantic of all sciences." Fields medalist Andrei Oukounov alludes to poetry when describing the work of a mathematician. As in writing poetry, he says, in writing pure mathematics, "you can distill it from one source and use it to infuse or illuminate anything you like." Only a poet could say something so cryptic about math.

Not all mathematicians have seen their field this way. Applied mathematicians sometimes scoff at the aesthetes who pursue pure math. Religious mathematicians have sometimes warned against the aesthetic attractions of the field. Beautiful math, they argued, might lead to sin. St. Augustine, for instance, once warned against mathematicians who followed in the footsteps of the Pythagoreans and thought that math ruled the universe: "The good Christian should beware of mathematicians and all those who make empty prophecies, especially when they tell the truth, for fear of leading his soul into error by consorting with demons." Mathematical beauty could beguile the gullible into thinking they know things that only God can know, and to fall into the original sin of pride that befell Adam when he ate of the Tree of Knowledge. In St. Augustine's time, "mathematician" was often synonymous with "conjuror," which, to Christians such as St. Augustine, was a satanic practice.

St. Augustine aside, many pure mathematicians throughout history have thought of their work as humbling in the face of the awesomeness of the universe and as essentially the same as the work of an artist who ennobles creation. Pure mathematicians do not solve scientific, economic, and political problems. They solve artistic problems. Applied mathematicians assess whether they have done good work by examining whether they have made something useful. But pure mathematicians assess whether they have done good work by examining whether they have made something beautiful. Hardy famously wrote, "Beauty is the first test; there is no permanent place in the world for ugly mathematics." Because a pure mathematician does not aim to do work that is useful, Hardy argued, the burden is even greater to make something that is lastingly beautiful.

The idea of evaluating math aesthetically might be confusing to a non-mathematician. Math is an academic field. It involves numbers, shapes, and measurements. We have all studied math in school. Arithmetic, algebra, geometry, maybe trigonometry, and even calculus. We were expected to get the right answers in these classes, not make something beautiful. As such, we might expect the criteria for judging math to be quantitative. At least we would expect those criteria to be standardized and rigorous. But beauty is qualitative. It is also notoriously subjective and loaded with cultural baggage. Does this mean that what makes "good math" in the field of pure math is up to each individual mathematician?

No, pure mathematicians say. Pure math is as rigorous and objective as applied math, only in different ways. Hardy's words substantiate this claim. "No other subject has such clear-cut or unanimously accepted standards, and the men who are remembered are almost always the men who merit it."

Hardy described those standards in his *Apology* using four key criteria. According to Hardy, beautiful math must be abstract. It must contain significant ideas, ideas that are deep but also general enough to connect different aspects of math. Beautiful math must be unexpected. It also must be intuitive. Once you know it, you must wonder how you didn't think of it earlier.

Math that fits these criteria—interconnected, abstract, intuitive, and unexpected—Hardy claimed, is not only beautiful but also *true*. These sensations help you understand the truth in a proof. They do not by themselves establish the validity of a proof. Only following the rules of mathematical logic can do that. A proof may be valid even if it is ugly. But if any of these four criteria are missing, you might have a hunch that the math you're looking at either isn't true or needs some additional work. Mathematicians will keep working on an ugly mathematical statement until it is not only logically valid but also beautiful.

Hardy outlined these criteria one hundred years ago, but mathematicians today still use them all the time. For a window into how they use these criteria, let's hear how one mathematician talks about his favorite mathematical idea: Burnside's Lemma.

MOHAMMED OMAR, a math professor at Harvey Mudd College, first learned about Burnside's Lemma when he was in college. He immediately loved it.

"I was like, Whaaaaat?!," he recalled to two mathematicians, Evelyn Lamb and Kevin Knudson, on their popular podcast *My Favorite Theorem*. Mathematician guests on the podcast talk about—you guessed it—their favorite theorems. To understand how much these mathematicians like their favorite theorems, you have to hear them talk about them. They *love* their favorite theorems. Remember how you felt the first time you heard your favorite song? That's how these mathematicians felt when they first encountered their favorite theorems. Like Omar, they reacted from the gut.

Significantly, the reasons they give for loving their favorite theorems often largely parallel Hardy's criteria for mathematical beauty. Even though their reactions to the theorems are emotional, there is a certain logic to their reactions. And the similarity of their reactions to each other and to Hardy's criteria support Hardy's contention, and the contention of other pure mathematicians, that Hardy's criteria for mathematical beauty are objective and valid.

Burnside's Lemma is a theorem you might run across when studying advanced math. It is named after a turn-of-the-twentieth-century English mathematician, William Burnside, who didn't actually develop it but merely wrote about it. A German mathematician named Frobenius seems to have developed it. (So it goes sometimes—fame can be fickle.) Burnside's Lemma unites subjects that most people would think have nothing to do with each other. Burnside's Lemma brings together three subjects: geometry, algebra, and combinatorics, the field of math entirely devoted to counting things. That combination is a source of its fascination.

When you first look at it, however, Burnside's Lemma doesn't seem particularly profound. In its practical application, it solves a rather pedestrian problem. A complicated problem, but not particularly earth-shaking. Basically, it helps mathematicians count patterns on a symmetrical object.

For instance, say we had a cube and wanted to paint its faces with three different colors: black, gray, and white. How many different ways could we do that? As Omar explains, we might first think about the cube having six faces, and we have three colors. So, each face could be painted one of three colors. That's $3 \times 3 \times 3 \times 3 \times 3 \times 3 = 729$. Easy peasy.

But that's not quite right. It's actually much less than that. We forgot about the cube's symmetry.

A cube is symmetrical. It looks the same from lots of different perspectives. Spin a solid white cube. How can we tell if the face we see is the same one we started with? Maybe we only did a partial spin. But maybe we did a complete three-hundred-and-sixty-degree turn. This is what mathematicians mean by *symmetrical.* An object is symmetrical if we can move it around and have a hard time telling that we did just by looking at it. The cube's highly symmetrical nature is what makes it such a good die. But it also makes it tricky to figure out how many different ways we could paint it with three colors.

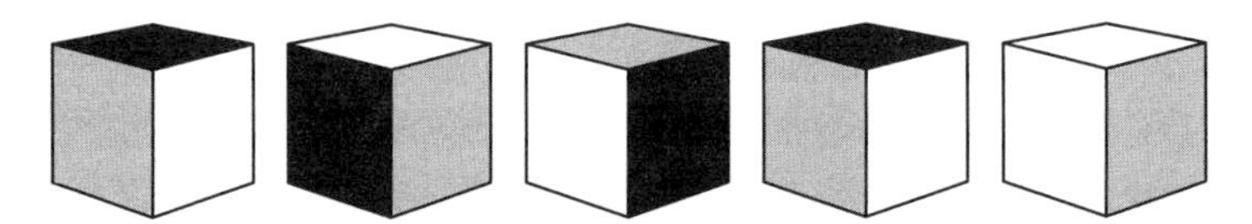

Here are five cubes with one black face, an adjacent gray face, and four white faces. (For the fifth cube, the face on the bottom is black.) How many such cubes can you find?

Let's try painting the cube. Paint the face facing you black, the face on top gray, and the remaining four faces white. That's the same as painting the face facing you black, the face to the left gray, and the remaining four faces white. To check, just spin the cube to put that new gray face on top. How many ways of coloring one face black, an adjacent face gray, and four faces white are the same if you spin the cube? If we want an accurate count of the number of different ways to paint the cube, we have to eliminate all of these identical combinations. For sure, 729 is too big a count. What might have initially seemed like a simple multiplication problem now turns into a tedious exercise in painting and spinning.

Here's where Burnside's Lemma comes to the rescue. Burnside's Lemma marshals the power of algebra to simplify this exhausting process. It might not be the kind of algebra you're familiar with. It doesn't involve solving for x. What it does involve is describing the structure of the symmetries of the cube. How many different ways can you spin a cube so that it looks the same? If you know all of those ways—that is, if you know the cube's symmetries—you're on your way to knowing which painting combinations to eliminate. Mathematicians call the set of an object's symmetries a *group*. Figuring out what spin moves are in a cube's group is much less tedious than trying to paint the cube in lots of different ways.

Burnside's Lemma says that if you want to figure out how many different ways there are to paint a cube with three colors, all you need to know is the structure of this group. And that's true for counting configurations on any symmetric object. Instead of counting and testing, counting and testing, you can use algebra to describe the structure of the object's symmetries. Algebra simplifies a problem that would be too tedious without it. To solve a counting problem on a symmetrical geometric object, you can just use some algebra.

So, that's nice. If for some reason you wanted to make painting a cube easier, you can use Burnside's Lemma. But does this make Burnside's Lemma beautiful?

Looked at from this practical point of view, as a way to simplify a painting problem, maybe this lemma is cute. It's cute like a group of kids singing "Head, Shoulders, Knees, and Toes" is cute. You might see them and think, "Awww, the song helps them learn their body parts." Similarly, you might see Burnside's Lemma and think, "Awww, Burnside's Lemma helps us paint cubes." But this certainly doesn't make Burnside's Lemma *beautiful*. Looked at in this way, it has none of the characteristics of mathematical beauty that Hardy described. It isn't particularly abstract. Or significant. It may be unexpected, but not in a moving way. And it isn't intuitive.

But Omar doesn't love Burnside's Lemma because it helps him solve counting problems. Omar's love affair with Burnside's Lemma stems from a different reaction. "My initial reaction," he said on the podcast, "it was a cool way to make some of this abstract math that we were learning really come to life." What does he mean by this? Omar sees in Burnside's Lemma a combination of mathematical characteristics that essentially meet Hardy's criteria for mathematical beauty.

First, Omar means that Burnside's Lemma connects fields of math that might otherwise seem far apart. Burnside's Lemma unites algebra, geometry, and combinatorics in a way that Omar finds satisfying. To Omar, then, Burnside's Lemma is what Hardy would call beautifully *significant*.

Next, Omar means that Burnside's Lemma takes algebra, a field that most often deals with abstract symbols, and gives it a visual, tangible application. It makes an abstract and sometimes confusing field more *intuitive*. It's hard to see and touch abstract symbols, which sometimes also makes them hard to understand. A cube captures the meaning of those symbols in an easy-to-handle form. Burnside's Lemma makes mathematics vivid and alive.

Finally, Omar means that Burnside's Lemma is simple and *abstract*, in that it captures a long counting process in a single algebraic object. And the combination of all these features in a single lemma is *unexpected*.

Significant, intuitive, abstract, and unexpected. These are characteristics of mathematics that Omar, Hardy, and other mathematicians love to see in tandem. They are decidedly detached from the lemma's uses. Burnside's Lemma doesn't help mathematicians solve problems that they particularly care about. Rather, their appreciation of Burnside's Lemma is related to the lemma's aesthetic qualities. Mathematicians agree: Burnside's Lemma is beautiful.

YOU MAY BE WONDERING, If Burnside's Lemma is an example of beautiful math, what is an example of *ugly* math? Hardy developed his set of standards for mathematical beauty so that he could distinguish between the two. Certainly, if there is beautiful math, there must be a counterpoint. There must be math that mathematicians acknowledge as logically valid but nonetheless disdain as unsatisfactory because it's ugly.

One of the most interesting things about math is that there is often more than one logically valid way to solve a math problem. It is one of the things that makes math a creative field. And it is one of the reasons that one can sometimes find both beautiful and ugly solutions to a problem. One of the best places to look for ugly math is among the incredible number of proofs of the Pythagorean Theorem. The Pythagorean Theorem might be the most famous mathematical statement. Mathematicians as far back as the ancient Greeks—and maybe further, if you believe some scholars of Plimpton 322—have known about the Pythagorean Theorem. Because it's so famous and has such a history, it makes sense that the Pythagorean Theorem would also be the most frequently proven theorem. But the sheer quantity of Pythagorean Theorem proofs is mind-boggling. In his book *The Pythagorean Proposition*, Elisha S. Loomis included no fewer than 371 different proofs of the Pythagorean Theorem, which could, for all we know, be just a drop in the bucket of Pythagorean Theorem proofs. It seems that everyone has a proof of the Pythagorean Theorem. Even US President James A. Garfield invented one.

Why would anyone want to read almost four hundred proofs of the same theorem? While all the proofs prove the same thing, they do so in different ways. The Pythagorean Theorem is to mathematics as *Hamlet* is to literature. Countless actors have portrayed Shakespeare's Hamlet. Producer after producer has reconceptualized the play. Similarly, mathematicians have returned to the Pythagorean Theorem again and again.

And mathematicians like some proofs of the Pythagorean Theorem better than others. Just as you can like some productions of *Hamlet* and dislike others, you can have taste in proofs. And your preferences may not necessarily be arbitrary. A production of *Hamlet* could be technically correct in that it includes all of Shakespeare's lines but still be generally disparaged as awful. Likewise, based on criteria akin to Hardy's guidelines for mathematical beauty, you can aesthetically judge a proof. On this basis, mathematicians have decided that some proofs of the Pythagorean Theorem are beautiful while others are downright ugly. So, let's look at a pair: a beautiful proof and its ugly stepsister.

Modern mathematician and spokesperson for mathematical beauty Steven Strogatz considers this to be his favorite proof of the Pythagorean Theorem:

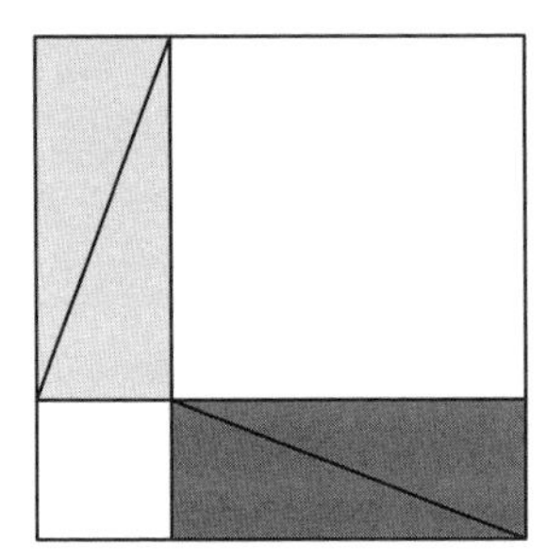

This may not look like a proof to you. It is just two simple pictures. It has no words, no numbers, not even any algebra. But it does prove the Pythagorean Theorem. In doing so, it hits many of the criteria for aesthetically attractive math that we've discussed. It's unexpected. It makes something abstract into something intuitive. It connects fields of math. And it feels inevitable.

Here's how it works.

The Pythagorean Theorem usually looks like $a^2 + b^2 = c^2$. That's the formula we all learned in school. But that's just an algebraic rendition of the theorem. The ancient Greeks didn't write it like that. The Pythagorean Theorem literally says that if you draw a square attached to the hypotenuse of a right triangle, using the hypotenuse as one of the square's sides, its area will always be the same as the combined areas of the squares drawn in the same way on the other two sides.

You can verify this yourself. Draw a huge right triangle. Any right triangle will do, but larger works better for this experiment. Then draw squares on all three sides of the triangle. Make sure that the squares use the sides of the triangle for their sides. Name the two smaller squares A and B, and the largest square on the hypotenuse C.

Then lay pennies, cereal, raisins, or any other small flat-ish objects side by side so that they completely fill squares A and B. Once you've filled the squares, move all the pennies, cereal, or raisins into the largest square, square C. Again, lay them side by side. See how close they come to filling square C? They won't fit perfectly. Filling a square with pennies is "real life," not math. Nothing in real life works as perfectly as it does in math. Which is one of the things that mathematical purists dislike about applied math. But the pennies should fit well enough to convince you that the C square—the square on the hypotenuse—holds as many pennies as do the A and B squares combined. Therefore the area of square C must be the same as the areas of squares A and B put together. As the Pythagorean Theorem says, the square on the hypotenuse is equal to the sum of the squares on the other two sides.

This penny experiment isn't a proof of the Pythagorean Theorem. Moving pennies from the smaller squares to the largest square gives us reason to believe that this might work all the time, no matter how big or small, skinny or fat the triangle. But it is inextricably tied to the particular triangle we started with. It isn't general enough to be a proof. It's intuitive, which is a good thing. But it isn't abstract. So, it isn't good enough to be true math.

Mathematician Steven Strogatz thinks that a truly beautiful proof of the Pythagorean Theorem should draw on our intuition about area in a way similar to moving the pennies around, and then build an abstract proof based on our intuition. We developed an intuition about the areas of the squares from moving pennies around. Our intuition helps us feel that the proof must be true. A beautiful proof of the Pythagorean Theorem should take advantage of that feeling. Strogatz's favorite Pythagorean Theorem proof does just that.

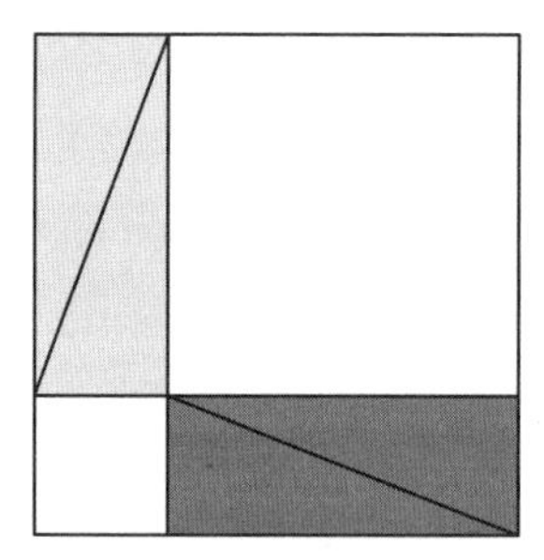

Let's start with the first picture. It is of a giant square made up of four right triangles, a small square, and a larger square. See how the two squares fit perfectly alongside the two non-hypotenuse sides of the right triangle? Those are our A and B squares from the penny experiment. The A and B squares, plus four right triangles, fit together to fill one larger square.

Now look at the second picture. It again shows a giant square. This square is the same size as the giant square in the first picture. But this time it is made up of the same four right triangles and a single, even larger, square.* Squares A and B are gone, replaced by a square whose sides are the hypotenuses of the four right triangles.

Recognize this square? It's square C! The C square, plus four right triangles, takes up the same amount of space as the A and B squares, plus the same four right triangles. So, square C must have the same area as squares A and B combined. In other words, A square plus B square equals C square.

Strogatz calls this proof elegant. He likes it because, as he writes, it "illuminates" the Pythagorean Theorem. His language about this proof is remarkably similar to Omar's about Burnside's Lemma—which, in turn, is remarkably similar to Hardy's. The proof is abstract. But it also draws on our intuition about how the Pythagorean Theorem relates to areas. It connects algebra with geometry. It's unexpectedly simple and straightforward. Once you understand what all of the shapes in the picture represent, all you need to do is rearrange the pieces, and you have a complete proof. You do not even need any words. Two simple pictures suffice. The proof is beautiful.

Ready for the ugly proof? Strogatz contrasts this beaut with the following, which he unabashedly calls "ugly." After presenting the ugly proof, he asks his readers, "Would you agree with me that, on aesthetic grounds, this proof is inferior to the first one?"

Let's see if we agree with Strogatz. If the standards for mathematical beauty are truly universal, then we should.

---

* I've left out an important part of the proof, which is how we know that we can arrange the triangles to make the two giant squares. This isn't too difficult to parse, though. The curious reader is encouraged to consider: How do we know that the two giant squares are the same size? How do we know that we can always build two such squares, starting with any right triangle?

Lay the right triangle down so that it rests on one of its non-hypotenuse sides (in this picture, side b).

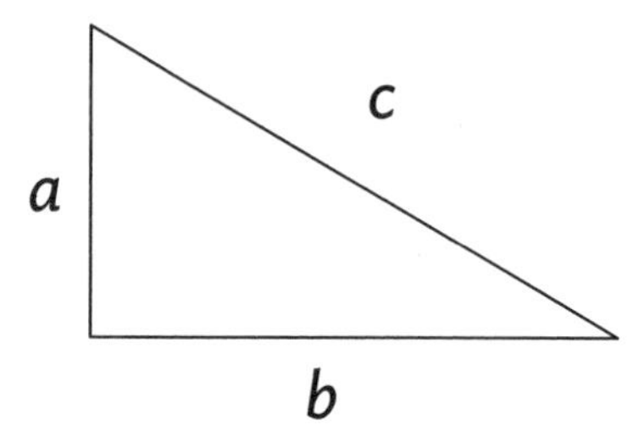

Then draw a line from the right angle all the way to the hypotenuse. Make sure that the line hits the hypotenuse in another right angle. You now have three triangles: the large triangle you started with, with sides a, b, and c, and two smaller triangles inside, one with sides a, x, and h, and the other with sides b, y, and h.

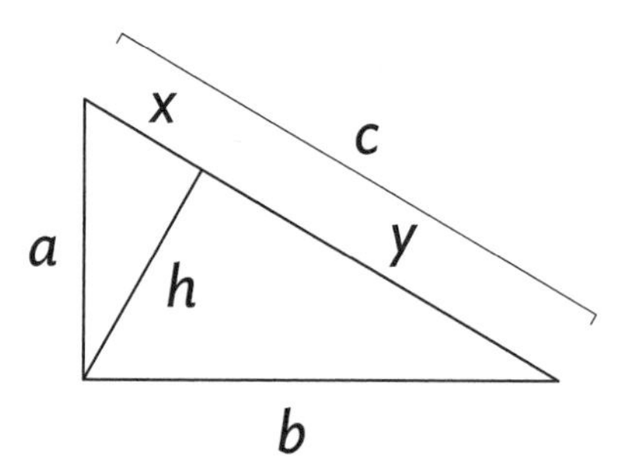

Now it's time to remember some geometry. You might remember that if you have several triangles that have angles of the same size, but the triangles themselves are of different sizes, then those triangles are *similar*. Saying that two triangles are similar means that they are the same basic shape. If you inflate the smaller triangles, you can make them exactly the same shape and size as the larger triangle. To make several copies of the same shape, all you have to do is scale the triangles up or down, like you're inflating or deflating.

The three triangles in this picture are similar. This is much easier to see if we separate the triangles and set them up beside each other, so that they look like a little family.

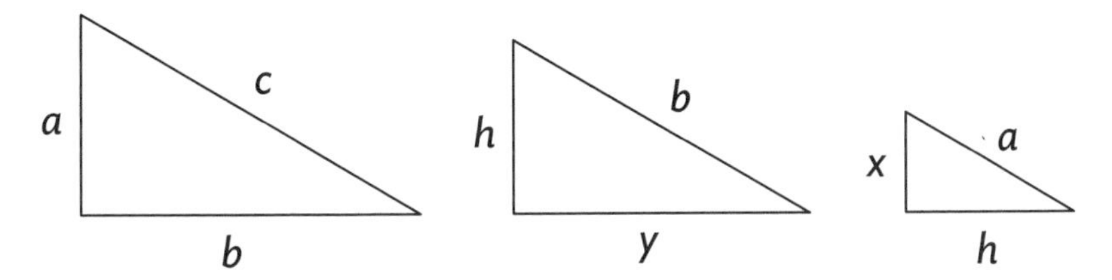

They certainly look similar, and a quick angle check shows that they are. Knowing this is the key to the proof we will construct.

Now we have to figure out how much to inflate the smaller triangles to turn them into the larger triangle. This process involves division. First, figure out which sides in the similar triangles match. These will correspond when you blow up the smaller triangle. The shortest leg in one triangle corresponds to the shortest leg in the other, and so on. Then, divide the lengths of the corresponding sides. You should get exactly the same number when you divide all of the corresponding pairs of sides. That's the number you'll scale by.

Now, if we divide the sides in the smallest and largest triangles in our little family, we find that

$$\frac{x}{a} = \frac{a}{c}.$$

That's because x and a, the short legs, and c and a, the hypotenuses, correspond to one another. Just because we can, and because it will be useful later, let's use some algebra to rearrange the equation above so that it has no fractions. If we cross-multiply, we have

$$a^2 = xc.$$

Now, let's divide the sides in the medium and large triangles:

$$\frac{y}{b} = \frac{b}{c}.$$

We can do this because y and b, the long legs, and b and c, the hypotenuses, correspond to one another. We'll use the same algebra to rearrange this equation, too:

$$b^2 = yc.$$

Believe it or not, we're getting close to finishing our proof of the Pythagorean Theorem. Only a little more algebra magic remains. Because we can, let's add together the two equations we made, $a^2 = xc$ and $b^2 = yc$. That gives us

$$a^2 + b^2 = xc + yc.$$

This looks more like the Pythagorean Theorem. There are some a's, b's, and c's around. But what can we do with the pesky x and y?

Look back at our original picture of the three triangles. See how x and y fit together to make c in the largest triangle? We can use algebra to express this by writing $x + y = c$. This little equation unlocks the end of the proof. Let the algebra wizardry begin.

First, we factor the c's to separate out the $x + y$:

$$a^2 + b^2 = c(x + y).$$

Then we substitute c for $x + y$, because we know they are equal to each other:

$$a^2 + b^2 = cc.$$

Finally, we remember that cc is the same thing as c^2 . . .

And, POOF! We've shown that $a^2 + b^2 = c^2$.

The Pythagorean Theorem appears, almost out of nowhere. At the end of a long, winding algebra road, we unexpectedly find ourselves at our destination.

What do you think? Is this proof ugly?

These two proofs of the Pythagorean Theorem, one that Strogatz thinks is beautiful and another that he thinks is ugly, end with the same result. And they are both logically valid. But they use different means to get there. They're as different as *Hamlet* and *The Lion King*. And the means they use will likely produce different reactions in the reader.

The first proof draws on an intuitive geometric interpretation of the Pythagorean Theorem. We can see, feel, and count the four triangles and three squares as they rearrange to fill both larger squares. Just as we saw, felt, and counted the pennies. The connection between the Pythagorean Theorem and area is intuitive. The result, produced with no words or algebra, speaks to us. This is why Steven Strogatz thinks it's beautiful. Hardy would almost certainly agree.

The second proof uses no squares. The operation of squaring doesn't even appear except as an algebraic afterthought, after we've taken a detour into the world of similar triangles. The similar triangles that the proof uses feel far away from the right triangles and squares that were used in the first proof. In the first proof, every time we say "equals," we mean "same size." That's intuitive. But in the second proof, it isn't clear what "equals" means. Sometimes it means: if I divide these pairs of side lengths, I should get the same number because of the laws of similar triangles. This makes sense. But other times it means: I know these things must be equal because I made them equal a long time ago and did a lot of algebra that shouldn't change their equality. Even if you are comfortable with algebra, this is much more confusing. If you aren't comfortable with algebra, it feels like magic.

Strogatz does not like the appearance of algebra in a proof of the Pythagorean Theorem. “Who invited all that algebra to the party?,” he asks. “This is supposed to be a geometry event.” Strogatz thinks that the algebra mucks up what could be a clear and clean geometric proof. The algebra takes something intuitive and makes it overly abstract. Abstract is good when it helps to generalize an idea, but it is bad when it makes the idea harder to comprehend. It makes the proof ugly. And, consequently, it makes the proof feel less true. He writes, “By the time you’re done slogging through it, you might believe the theorem (grudgingly), but you still might not *see* why it’s true.”

As a result, the second proof isn’t just ugly. It’s also bad math. Its ugliness has implications for its mathematical value. It doesn’t do a good job of convincing, Strogatz argues, because of its ugliness.

I agree with Strogatz—mostly. I, too, enjoy the first proof for the ways in which it draws on my intuition. And I dislike the second proof because it sacrifices intuition to abstraction. But I don’t think the second proof lacks mathematical value. I don’t think it’s completely ugly. I see something attractive in the second proof that the first proof doesn’t have: the element of surprise.

Unless you already know how the second proof works, you don’t see the Pythagorean Theorem coming. It pops out of the algebra at the very end. The magical quality of the proof, which undercuts its intuitiveness, contributes at the same time to its unexpectedness. The fact that you can derive the Pythagorean Theorem using similarity is surprising. It’s unexpected. It’s also significant. Similarity relationships among right triangles are important in math, especially in trigonometry, in which the Pythagorean Theorem also plays a starring role. I love how the second proof makes an unexpected connection between the Pythagorean Theorem and similarity relationships in right triangles.

But wait—connections to other branches of math and unexpectedness are two of Hardy’s criteria for mathematical beauty. If the second proof has these features, how can it be unconditionally ugly? Yes, the first proof fulfills all four of Hardy's criteria, but the second proof is clearly more surprising. How should we weigh the relative strength and weakness of the four criteria in a proof?

In these two proofs, we begin to see some tension within Hardy's criteria. Sometimes math can't be simultaneously interconnected, abstract, intuitive, and unexpected. The unexpected abstraction and interconnectedness in the second proof come at the cost of intuition. Maybe it's off-putting. Sometimes the unexpected is jarring. But it helps you learn something new, so maybe it isn't completely ugly.

There are plenty of people who would prefer to never be taken by surprise by a piece of mathematics. We like our mathematics to be logical and orderly, these people say. Please only show us mathematics that makes immediate sense. But math is full of surprising results. For instance, I think that the Twin Primes Conjecture—that there are infinitely many prime numbers that are two apart—is surprising. Prime numbers go on forever and usually get further apart as they get larger. And yet the Twin Primes Conjecture says that no matter how high you count, you will always be able to find a pair of primes that are two numbers apart. To me, that is as disconcerting as it is surprising. But that doesn't mean it's *ugly*. The mathematician in me is happier knowing this surprising conjecture.

Unexpected math takes you by surprise. But don't resent the math for it. Surprise is how you learn.

In my opinion, neither of these two proofs of the Pythagorean Theorem reaches the pinnacle of mathematical beauty. Neither has the blend of intuition, connectedness, abstraction, and unexpectedness that appeals to me most. Strogatz's proof is undeniably beautiful, albeit a bit short on surprise. The second proof may be on the uglier side, but it has some beauty in its unexpectedness. And they both teach me something new. That is key to the development of math, to keep improving on what we have already proved. Just as each new rendition of *Hamlet* adds something to the Shakespearean canon and helps us better understand the human condition.

I do have a current favorite Pythagorean Theorem proof, however. It is my favorite because I think that it balances intuition, abstraction, connectedness, and surprise in a beautiful way. Can it be improved? Maybe. Maybe you'll find a way.

This proof comes to us from Ricardo Pérez-Marco, a mathematician at the University of California, Los Angeles. It uses the geometric notion of squaring so perfectly captured in the first proof. And it uses similarity relationships in right triangles, which the second proof used, but it does so without unnecessary abstraction and distraction. In my opinion, it has the best of both worlds.

Not everyone is as struck by this proof as I am. One of Pérez-Marco's colleagues, mathematician Terence Tao, calls the Pérez-Marco proof "very cute" but "not particularly earth-shattering." So, mathematicians can disagree on the relative beauty of a proof. That is the nature of art and the art of mathematics. But he adds that it "is perhaps the most intuitive proof of the theorem that I have seen yet." It is its intuitiveness that particularly impresses me about the proof. Read the proof and see what you think.

I've left out some of the connections in the proof. You will have to make those connections yourself in order to fully finish the proof. It might be hard, but give it a try. I think that making those connections yourself is part of the aesthetic experience. Enjoy!

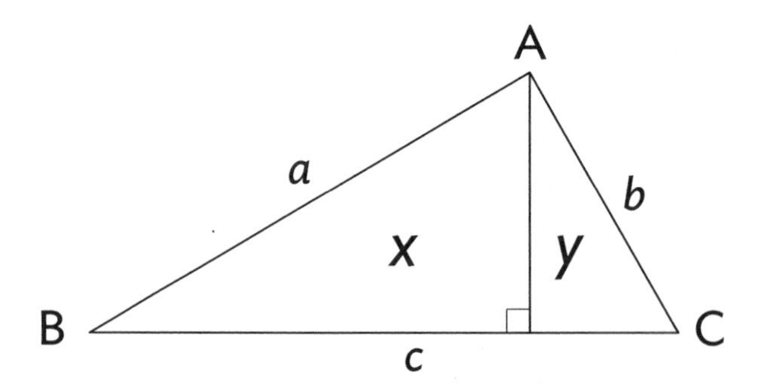

"In the diagram opposite, a, b, c are the lengths BC, CA, and AB of the right-angled triangle ACB, while x and y are the areas of the right-angled triangles CDB and ADC respectively. Thus the whole triangle ACB has area $x + y$.

"Now observe that the right-angled triangles CDB, ADC, and ACB are all similar [because of all the common angles] and thus their areas are proportional to the square of their respective hypotenuses.* In other words, $(x, y, x + y)$ is proportional to (a^2, b^2, c^2). Pythagoras' theorem follows."

Isn't that nice? It gives us intuitive squares (which we have to imagine in the proof's current rendition, but they're still there). It also gives us connections to similar triangles and an unexpected leap to the Pythagorean Theorem. I like it.

The Ugly Duckling

Hardy's criteria provide mathematicians with an objective way of evaluating mathematical beauty. We have seen how those criteria help us understand what mathematicians value in Burnside's Lemma and how mathematicians choose between proofs of the Pythagorean Theorem. But although Hardy's criteria help mathematicians evaluate their work, that does not mean that mathematicians will always agree on which math is or is not beautiful. Objective does not necessarily mean "exact." Objective means that there are generally agreed-upon standards for deciding a question

* This part of the proof might be the hardest to understand. Here's a hint: the areas of the triangles are in proportion because the triangles are similar—that's what "similar" means. The areas of the triangles can't be proportional to the lengths of their hypotenuses, though, because the dimensions of the quantities being compared are different. Comparing length to area is like comparing apples to oranges, or so the saying goes. But if you *square* the hypotenuses, you take care of the dimension problem.

and that you consider all the plausible answers to the question. And you can provide reasons for choosing your answer.

Many mathematicians agree on Hardy's list of criteria for mathematical beauty. And those criteria provide an objective means of evaluating math aesthetics. But sometimes they seem to conflict. Surprise, for example, can conflict with intuition. Intuition is comforting. Surprise is disconcerting. In an otherwise logical proof, it can feel unnatural. Even false.

You might think that mathematicians' squabbles about mathematical beauty are harmless. Does it matter that they don't always share the same aesthetic reactions to surprising math? To some extent, it doesn't. It matters just as little as if you and I disagreed about whether we liked a particular painting. We are each entitled to our own preferences.

But standards of mathematical beauty play an important role in arbitrating what math is good and what math is bad. Mathematical truth is tethered to beauty. Hardy's criteria do not merely express matters of subjective preference. Remember, Hardy said, "No other subject has such clear-cut or unanimously accepted standards, and the men who are remembered are almost always the men who merit it." Hardy's criteria describe objective criteria that mathematicians should consider when evaluating their work. They ignore his criteria at their own peril. And sometimes they do. Of Hardy's four criteria—abstraction, interconnection, intuition, and surprise—surprise is often ignored and rejected. And that can lead to important mistakes in pure math, including mistakes that can have ramifications in applied math.

We all have a tendency to go with the tried and true, even when it might no longer be entirely true. Even mathematicians suffer from this tendency. But if we reject surprise because it is disconcerting, and maybe reject a new idea because it conflicts with ideas we particularly like, we close ourselves off from the possibility of progress and run the risk of wallowing in error. This is true in any field, including mathematics.

The recent history of mathematics has at least one story in which that is precisely what happened. Mathematicians were so blinded by their attachment to a form of intuitive and conventional geometry that they missed an entirely new world. They thought the new world was ugly. They were suspicious of it because it didn't fit with their conventional conceptions for beautiful math. And they rejected as illogical and ugly a new form of math that would eventually revolutionize what mathematicians think of as logical and beautiful. The ugly duckling turned out to be a swan.

THAT MATHEMATICAL UGLY DUCKLING is called hyperbolic geometry. It was developed in the mid-nineteenth century by a mathematician named Janos Bolyai. He was bold enough to reimagine what mathematicians meant by intuitiveness. In doing so, he built a mathematical world that contradicted established mathematical knowledge and mathematicians' intuitive sense of the nature of space.

Bolyai's mathematician colleagues were terrified of hyperbolic geometry. And they weren't completely wrong to be. Hyperbolic geometry is weird. And it's a little scary.

If your only exposure to geometry was in high school, you've probably heard of only one type of geometry: Euclidean geometry. Euclid was an ancient Greek mathematician who laid down the basic rules of this type of geometry. Those rules have been used by mathematicians for two thousand years. At one time, they were considered almost sacred. Euclidean geometry is geometry that takes place on flat spaces. Infinitely long and wide pieces of paper and infinitely long, wide, and tall boxes are the settings of Euclidean geometry. Euclidean geometry also takes place in our daily lives, as we move things around on tables, peer up at tall trees, and travel short distances. And without Euclidean geometry, we wouldn't have things like the Pythagorean Theorem. Euclidean geometry is useful and largely represents how we see the world.

But there are other types of geometry. You may also have heard of spherical geometry. If you haven't heard of it, you've certainly experienced it. Spherical geometry takes place, as you may have guessed, on the surface of a sphere. Spherical geometry governs what happens when we travel long distances on Earth. There are obvious differences between spherical geometry and flat-plane geometry. One of the most important is that if you move in a straight line on a sphere, you'll actually travel in a giant circle. Have you noticed how the shortest flight between two distant cities seems to curve on a flat map? That's because the cities are on a sphere. What looks like the shortest flight between the cities on a flat map isn't really the shortest.

Spherical geometry may be different from Euclidean geometry, but the two are not inconsistent with how we understand the world. Spherical geometry is not disconcerting. Hyperbolic geometry is.

You might not have heard of hyperbolic geometry. It is not taught in high school math classes. It is not an obvious part of our everyday lives. It wasn't even a part of mathematics at all until Janos Bolyai dared to suggest it.

In a nutshell, hyperbolic geometry is the geometry of ruffled space. It feels similar to Euclidean and spherical geometry if you only travel short distances. You might not notice that you lived on a hyperbolic plane if you never left your hometown. But if you traveled far enough, things would start to get strange.

Hyperbolic space constantly expands under your feet. We take for granted in flat space and on a sphere that if two people start walking in the same direction from the same initial spot, they'll end up at the same destination. How else would we give directions? But this isn't true in hyperbolic space. In hyperbolic space, infinitely many people can walk in the same direction from the same initial spot and end up in infinitely many different places. You wouldn't want to get on an airplane traveling in hyperbolic space. No one would know where you'd land.

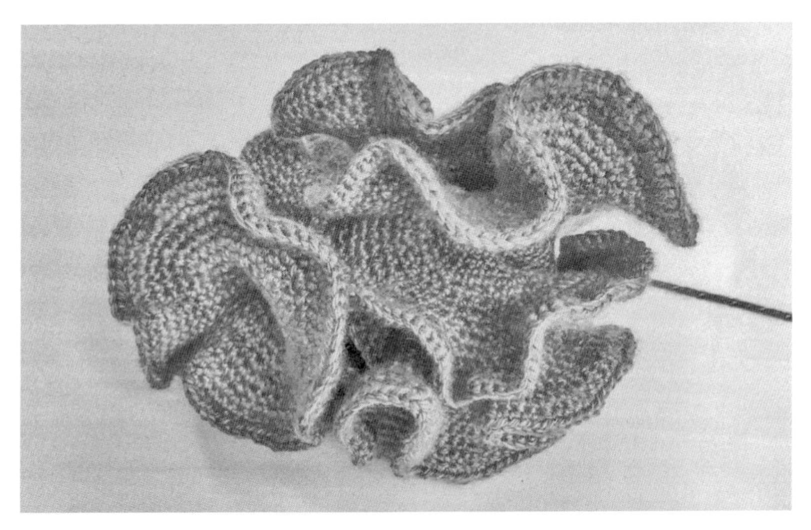

This model of hyperbolic space, an approximation of something called a hyperbolic plane, was crocheted by mathematical artist and educator Malke Rosenfeld.

Hyperbolic space sounds bizarre. It sounds like it shouldn't exist. But it does exist in our world. Lettuce leaves and pieces of coral are hyperbolic spaces. Despite these examples, of real-world things that we encounter in our kitchens and on our beaches, mathematicians before the nineteenth century didn't know, and didn't want to know, that hyperbolic geometry existed. They feared it.

Hyperbolic geometry didn't terrify mathematicians just because it was weird. They feared it because it violated one of the most basic tenets of Euclidean geometry, which was supposed to be perfect. After all, it came from the most beautiful book about mathematics ever written, Euclid's book *The Elements.*

Mathematicians for thousands of years after Euclid considered *The Elements* to be math in its ideal, most aesthetically pure form. They believed that his comprehensive mathematics described deep truths about the world around them. Given how the book is organized, you can see why. Euclid's *Elements* is a perfect example of Hardy's criteria for mathematical beauty. In this book, Euclid uses a small set of intuitive basic rules, which mathematicians call postulates, to prove and connect abstract geometric ideas about lines, angles, shapes, and other two- and three-dimensional objects. The way in which he builds complex math from such simple beginnings is unexpected.

The Elements was the mathematicians' Bible. It endured for thousands of years because it was not only logical and useful, but also beautiful. Something so clean and well ordered, which grew so naturally without hiccups, must be true. Unfortunately, the narrow-minded application of those qualities also led mathematicians astray.

Remember how walking in the same direction in hyperbolic space can take different people to different places? The fifth postulate that Euclid introduces in *The Elements* says that this is not possible. This postulate is called the parallel postulate. It essentially says that if two people start off walking parallel to each other, they will always stay parallel to each other. They'll be the same distance from each other when they start walking as they were when they started. For centuries, mathematicians took Euclid's parallel postulate to be gospel truth and tried to prove it. The idea of hyperbolic space was blasphemy against Euclid.

But they were wrong. It is logical to have a space where the parallel postulate is true. It's also logical to have a space where the parallel postulate isn't true—namely, in hyperbolic space. That's why the parallel postulate can be postulated but not proven. But it was important to mathematicians before Janos Bolyai to prove that hyperbolic space was logically impossible. They tried to prove it because they thought, on aesthetic grounds, that they could. They thought it wasn't a postulate. They thought it was a theorem.

To understand why mathematicians were so upset, we have to learn more about the differences between postulates and theorems. A *postulate* is an intuitive, basic rule. Postulates are assumptions. Mathematicians make them because they feel obvious. Postulates must be so intuitive that you don't feel the need to prove them. To be convincing as a postulate, then, a mathematical statement must have a particular aesthetic feel to it. It should also be simple to state and see. It shouldn't take too many words to explain or make you think too long before you understand.

All subsequent mathematical statements are proven using postulates as their building blocks. These subsequent statements are called *theorems* (or, sometimes, lemmas, as in Burnside's Lemma). Mathematicians don't expect theorems to feel as intuitive as postulates. That's because mathematicians do logical deductions using intuitive postulates to show that theorems are true. So, aesthetically, theorems can be less intuitive, more complex, and wordier.

In mathematics, postulates carry a big aesthetic burden. Not only do they have to assemble nicely using logic to prove the theorems that follow them, but they also have to strike mathematicians as intuited, rather than deduced. They are the foundations on which all further mathematical discoveries are made.

In *The Elements*, Euclid attempted to build geometry by declaring a minimal number of postulates. He did as much as he could with his first four postulates. Aesthetically, they all feel like postulates. They are easy to describe, intuitive, and, arguably, beautiful. They fit mathematicians' aesthetic expectations for what postulates should be like.

Take the first postulate, for example: "Each pair of points can be joined by one and only one line." This is obvious. Try to draw two points and see if you can make more than one straight line connect them. No way. The second says that a straight line can be extended indefinitely. Seems sensible. Or the third: "If you have two lines of different lengths, you can cut off a piece from the longer one that's equal in length to the shorter one." Nothing controversial there. The fourth is that right angles are all the same size. That's obvious enough. A right angle is a right angle is a right angle.

The first four postulates seem perfectly obvious, intuitive, and universal. And Euclid used his first four postulates to prove many theorems. Eventually, however, he needed to introduce a fifth postulate, the infamous parallel postulate. Unfortunately, the fifth postulate is nowhere near as postulate-like as the first four. It's not nearly as intuitive. And it's decidedly not beautiful.

Here's Euclid's statement of the fifth postulate:

> If a straight line falling on two straight lines makes the interior angles on the same side less than two right angles, the two straight lines, if extended indefinitely, meet on that side on which the angles are less than two right angles.

If your first reaction to this postulate is to scratch your head, you're not alone. It's almost as confusing as it would be if I'd given it to you in ancient Greek. It's hard to fathom what Euclid meant by this postulate. Even if you try to draw a picture, his conclusion is not obvious. This postulate is much less intuitive than the first four. It feels inappropriate to assume something so unobvious.

We can thank late eighteenth- and early nineteenth-century British mathematician John Playfair for popularizing a much simpler wording of the fifth postulate. You may be more familiar with this one: "Through any point in the plane, there is at most one straight line parallel to a given straight line." Oh, *now* that sounds like a parallel postulate.

Playfair showed that this is equivalent to Euclid's confusing phrasing of the fifth postulate. But while this wording is much easier to understand, it is still far less intuitively true than are the other four postulates. The other four are so intuitive that they hardly feel worth mentioning. The fifth, though—you might need some convincing before you can take this one for granted. And if you need convincing, maybe you don't think this should be a postulate. It might feel more like a theorem.

Many mathematicians after Euclid felt precisely this way. They were convinced on aesthetic grounds that the fifth postulate was actually a theorem. They thought it should be proven using the first four postulates, not posed as a postulate that would be accepted on the basis of intuition. Some mathematicians felt so strongly about this that they considered anyone who accepted the fifth postulate as a postulate to be stupid. For instance, Proclus, another ancient Greek mathematician, snidely commented that those who think the fifth postulate is intuitive are "yielding to mere plausible imaginings"—a serious insult for any serious mathematician. So, while mathematicians accepted Euclid's first four postulates without qualms, they were desperate to find a more solid foundation for the fifth. They set out to prove it as a theorem, using the first four postulates for support.

Mathematicians tried to prove that the parallel postulate was a theorem for two thousand years. Many of the most famous mathematicians worked on it. No one succeeded. As you might imagine, they started to get desperate. Mathematical aesthetics had never before let them down. Euclid's beautiful system seemed in danger. The parallel postulate simply could not be a postulate. But no route to proving it as a theorem got them to their desired destination.

Things came to a head at the beginning of the nineteenth century. One of the greatest mathematicians of all time, Carl Friedrich Gauss, was banging his head against a wall trying to prove the fifth postulate as a theorem. He began to doubt the validity of *The Elements*, that holy mathematical text. He wrote to Farkas Bolyai, Janos's father, of his attempts to prove the parallel postulate: "In my own work thereon I myself have advanced far (though my other wholly heterogeneous employments leave me little time therefor), but the way, which I have hit upon, leads not so much to the goal which one wishes, as much more to making doubtful the truth of geometry." If Gauss was growing doubtful of the truth of geometry, which for him and everyone else was synonymous with Euclid's geometry, things were getting really bad.

Spurred on by the need to protect Euclid's work, Farkas Bolyai also dove into the dismal abyss of trying to prove the fifth postulate. The work made him miserable. He wrote that "every light and every joy of my life has been extinguished" by it. But he cared so much for the integrity of Euclid's work that he was willing to martyr himself for the cause. He wrote, "I had purposed to sacrifice myself to the truth; I would have been prepared to be martyr if only I could have delivered to the human race a geometry cleansed of this blot."

Imagine how Farkas felt, then, when he learned that his son was treading the same path. He begged Janos to stop. "For God's sake, please give it up," he wrote. "Fear it no less than the sensual passion, because it, too, may take up all your time and deprive you of your health, peace of mind and happiness in life." Has anything more nineteenth century ever been written?

But Janos didn't give it up.

Maybe Janos was reckless. Maybe he thought his father Farkas was being overly dramatic. Maybe he happened to buy a head of lettuce or go scuba-diving in the Great Barrier Reef.

In any case, in 1823, twenty-year-old Janos took on the parallel postulate himself. But he took a different approach. Instead of assuming that the only universe is one in which the parallel postulate holds true as a theorem, Janos imagined the opposite. What if Euclid's first four postulates could be true without the fifth? What would a universe be like in which there could be more than one straight line through a single point parallel to a given straight line?

This was a good strategy. Assuming the opposite and finding a contradiction is a tried-and-true way that mathematicians prove things. If you explore all possible consequences of the opposite of something that you hope to be true, you might reach a ridiculous and clearly false conclusion. Then, you'll have shown that this opposite assumption must also be false. Tuliya and Jaden took this approach in their proof that there are infinitely many primes. But there can be pitfalls in this method. Assuming the opposite can dangerous. The opposite is messy, confusing, and ugly. Worse yet, the opposite of what you are trying to prove could itself be true. But mathematicians who take the risk of wading through the swamp of opposites often find the result they were hoping for.

Janos, however, didn't. He found something completely different. He didn't find what he, his father, Gauss, and countless other mathematicians had hoped for. Contrary to the contention of the fifth postulate, assuming that there can be many straight lines running through one point that are parallel to another straight line does not end in a contradiction. It leads you to a new kind of space. That ruffled, ever-expanding, lettuce-like space that does not fit properly in our everyday world but exists nonetheless. Hyperbolic space.

This is crazy and counterintuitive. Maybe it's even in some sense ugly. But it's true. Points and lines can behave as we would expect them to under Euclid's first four postulates and still exist in a universe as bizarre as hyperbolic space. Janos Bolyai had realized that Euclid's geometry was not the only one. He wrote to his father in 1823, "I created a new, different world out of nothing." He published his new geometry in an appendix to Farkas Bolyai's treatise on geometry, *Tentamen.* Now, almost two hundred years later, Janos Bolyai's appendix is what we remember, not Farkas's major treatise.

That's because mathematical aesthetics had led mathematicians astray for more than two thousand years. It took a mathematician willing to break the mold to see that beautiful math did not have to be bound by Euclid. The ugly duckling turned out to be a beautiful swan. And Bolyai's bold finding—which was shared by a handful of other mathematicians working at the same time—opened up huge new fields of mathematical research. It has also presented opportunities for mathematicians to try to create new forms of mathematical beauty. Would Eve Torrence's *Day* and *Night* sculptures, made from hyperbolic planes, have been possible without it?

Janos Bolyai's discovery of hyperbolic space shows that the rules of mathematical aesthetics may be shared, but their application might not always be uniform. This is not a problem that is unique to math. In every field of knowledge, from the most pure to the most applied, there are principles and applications. People regularly disagree about them.

While it is a problem when people don't agree about important things, it is also an opportunity to find new solutions to problems. Despite our conventional school math assumption that the solution to a math problem must be definite, and not up for debate, math is just like other subjects in being simultaneously definite *and* debatable. For instance, what does intuitive *really* mean? Before Janos Bolyai, nothing could have been less intuitive than hyperbolic space. But now that mathematicians know about hyperbolic space, they find it everywhere, not just in lettuce and corals. In fact, physicists think that our universe might behave more like hyperbolic space than Euclidean space. What we think of as intuitive is limited to our experience. But when we expand that experience, our intuition grows—even to include worlds as strange as hyperbolic space.

Hyperbolic space may have felt unintuitive, ugly, and therefore invalid to mathematicians before Janos Bolyai. But once we spend some time understanding hyperbolic space and uncovering parts of our world that follow its strange rules, we realize that hyperbolic space is beautiful, too.

An Unlikely Mathematician

What does the story of Janos Bolyai's ugly duckling geometry mean for Hardy's standards for mathematical beauty? For applied mathematicians, the standard of determining excellence is the usefulness of their math. But pure mathematicians need a different standard. Mathematicians before and after Hardy have relied on the standards for beauty that he described. Those standards have led them to produce meaningful mathematical work. But a narrow, mechanical reliance on standards may hinder them from learning new things, especially things that unpleasantly surprise cherished ideas.

This is a problem. We're used to math solving the problems. But maybe this time math needs to look elsewhere to solve its own problems. Perhaps it should look to art.

There are standards for artistic beauty, just as there are standards for mathematical beauty. But as philosophers of mathematics and aesthetics Nathalie Sinclair and David Pimm observe, standards for beauty in art have changed dramatically through the millennia. And innovations in art have followed. Artistic innovators have forced changes in artistic standards, or at least what is considered acceptable under the existing standards. The same could be the case with math.

The arts, however, attract larger numbers of people than does math, both as practitioners and patrons. Sinclair and Pimm also observe that mathematics attracts only a small group of "like-minded people." There is not the critical mass of mathematicians and math fans to produce the almost continuous stream of innovators that we see in most arts. Innovative artists and "aesthetic revolutionaries such as Picasso, Pollock or Cage do not have mathematical equivalents." Janos Bolyai's story illustrates this situation. Styles and tastes in math largely did not change in the two thousand years after Euclid. Even today, long after mathematicians have recognized hyperbolic space as valid, Euclid's *Elements* is often still considered the pinnacle of mathematical beauty. When looking for a beautiful proof in the early 1900s to share with his readers of his book, Hardy didn't choose a modern piece of mathematics. He chose one of Euclid's.

Sinclair and Pimm wonder why this is the case. They also wonder, How can we make math more like art? How can we attract a mathematical Picasso, while also maintaining the logical and objective structure so important to the field?

Rejecting criteria for mathematical beauty altogether isn't a good solution. An "anything goes" approach to the value of mathematics, beyond just whether or not it's logically sound, would conflict with the desire for mathematical rigor. If you don't have any aesthetic standards to apply, you can't decide whether a piece of mathematics has value beyond whether it's logically sound. Mathematicians care deeply about the aesthetic value of their work. New criteria to replace Hardy's might, in turn, be useful, but no one has yet proposed anything that has gained widespread acceptance. The beauty of Hardy's criteria, and a main reason they have survived so long, is that they are broad and can accommodate many different applications.

Maybe the problem is less with the criteria themselves than in how they are applied. And, in particular, how they are used to weed out or discourage new mathematicians with radical ideas.

Only a small number of people have access to making or even experiencing beautifully innovative mathematics—of either the pure or applied varieties. As far back as elementary school, people experience math as rules to follow, not a space in which they can create. School math isn't full of sunny paper polyhedra or geometric spaces that defy the imagination. It isn't even full of problems that feel relevant to people's lives.

Students rarely have the opportunity to make meaningful choices in mathematics. Problems aren't posed to students in ways that invite creativity. Schools rarely say, "Here's a curious situation with two fractions. How would you approach it?" Instead, they typically say, "Here's what you do with two fractions. Now do it. And then do it again. And again." Students are told how to solve the problems before the problems are even given to them. Students' interaction with math is less like problem-solving than like practicing a technique.

For the lucky few students who do get to have aesthetically satisfying experience with math, math is attractive. For those who don't, math can be off-putting, even repulsive.

But Nathalie Sinclair has demonstrated that school-age children are capable of making aesthetically driven claims about mathematics. Her studies also show that when students have the opportunity to make decisions about math guided by their tastes, opinions, and sense of beauty, they thrive in their math classes. Children might not make the same choices as the great mathematicians who came before them. And children have a lot to learn about how mathematicians work. But children who are allowed to express their different tastes might have a stronger engagement in mathematical learning. They might ultimately help to advance the field.

And what about adults? What about adults who make innovative, groundbreaking work, despite the stringent standards of the field? They exist—even if their work doesn't always get its moment in the sun. You might never have heard of Eve Torrence, BEAM students Tuliya and Jaden, Nicky Case, Julia Angwin, or the Oksapmin before opening this book. They don't work at prominent research universities. They haven't published peer-reviewed papers in the most famous mathematical journals. But their work is groundbreaking. It makes us rethink our assumptions, both about mathematics and about the world around us. It is unquestionably beautiful.

The creativity of these mathematical outsiders would not be possible without the support of mathematical insiders who are willing to broaden their own horizons and invite someone new to join the insiders' club. But it's not enough to admit people with innovative ideas only to pressure them to conform to existing standards and structures. What people with new ideas need most is a real seat at the table and a chance to build new structures for themselves. When mathematicians are willing to watch their standards evolve in the hands of those with new ideas, the field grows.

There is perhaps no more touching illustration of this principle than the story of Marjorie Rice.

MARJORIE RICE HAD AN unremarkable childhood. She was born in Florida in 1923. As a high school student, she was tracked into a training program for secretaries.

For a few years she worked in a laundry and a printing office. Then, at the age of twenty-two, she married her husband, Gilbert. Like many women of her time, Rice spent the next thirty years caring for her family. She took their five children to school and the library. She made breakfast, lunch, and dinner for her family in their sunny San Diego kitchen with pea-green appliances.

If preparing meals was all she had done in that cheerful kitchen, Rice's life might have gone on just as it had started, full of friends, home, and family. Happy, but unremarkable. But in 1975, at her kitchen table, Rice did something completely different. Something remarkable. She made a mathematical discovery that mathematicians had neither seen nor anticipated before.

Rice discovered a new type of pentagon that could tile flat space. By 1977, she had discovered three more. This may not sound important. But in doing so, Rice blew open a branch of mathematics that professional mathematicians had previously considered closed and inspired a chain of mathematical research that was resolved only in 2017.

The question of which polygons can be used like tiles to cover a flat space has puzzled mathematicians for thousands of years. Mathematicians call questions of this sort problems of tiling the plane. It's pretty easy to see how some polygons tile the plane. You could probably come up with a good list of polygonal tiles if you tried right now. All triangles and quadrilaterals, three-sided and four-sided shapes, tile the plane, no matter how irregular their side lengths and angle sizes. That is, you could cover a floor with copies of the same triangle or quadrilateral and leave no gaps. But shapes with more than six sides make terrible floor tiles. If you drew a few pictures, you could probably prove to yourself that if a polygon has more than six sides and is convex—meaning that all of its angles are smaller than one hundred and eighty degrees—it cannot possibly tile the plane. It's just too big to fit together with others like itself.

But it's not so obvious whether and how you can tile five- and six-sided polygons, also known as pentagons and hexagons. Take convex pentagons, for instance. Pentagons are tricky to use as tiles. This is in part because a regular pentagon, the most basic pentagon with all equal sides and angles, does not tile the plane. The pentagon shaped like a house, with a square base and triangle top, does. But it's clear that not all pentagons tile the plane. So, figuring out which ones do is much more difficult than the same task with triangles and quadrilaterals.

Are there just a handful of pentagons that tile the plane? Or are there types of pentagons that tile in similar ways? And how would you ever know that you'd found them all? There are infinitely many pentagons. You could be lured into thinking you'd found them all simply because of lack of imagination. But a sneaky tiling pentagon could still be hiding from you.

This is precisely what happened to Johns Hopkins University mathematician Richard Brandon Kershner in 1967. Kershner published a paper in the prestigious *American Mathematical Monthly* claiming that he had found all possible categories of convex pentagon tiles. Kershner's findings made a big splash in the world of mathematics. So big that one of the most widely read mathematical writers of the time, Martin Gardner, wrote about them in his famous *Scientific American* column.

Gardner's column traveled to schools, offices, libraries, and, in Rice's case, kitchen tables all over the country, sharing the good news. Finally, a mathematician had found all of the convex pentagons that could tile the plane.

But Kershner was wrong. He hadn't found all of the convex pentagons that tile the plane. Not by a long shot. Kershner said that there were eight types of tiling convex pentagons. Mathematicians now know that there are fifteen. And it was Marjorie Rice who first proved him wrong, sitting at her kitchen table in the glow of those pea-green appliances.

Rice had been sneaking peeks at her youngest son's copies of *Scientific American* for years. She loved reading Gardner's column. She had always had a soft spot for math, but she hadn't taken more than one course in high school. Math simply wasn't available to her in the track on which she had been set. But while her children were away at school, Rice had some time to herself to learn math from Gardner.

The article on pentagonal tilings grabbed her interest. "I first discovered my interest in this when Martin Gardner wrote about the eight types of pentagon tiling in his column," she said in an interview in 1996. "And I thought, my, that must be wonderful that someone could discover these things which nobody had seen before, these beautiful patterns."

Spurred by an aesthetic appreciation of tiles and a competitive curiosity, Rice began to noodle around, looking for a new convex pentagon that tiled. She had no formal mathematical education, so she developed her own notation to keep track of the complex mathematical work she was doing. After a few months, Rice had found a new convex pentagonal tiling of her own. That's no time at all in the world of mathematical discoveries, especially those that refute published work done by professionals.

"I was so delighted when I found it—I couldn't believe it," she said. "I didn't really think I would find anything. I just was so thrilled. And I sent it to Martin Gardner."

Martin Gardner must have gotten many letters from readers with "new mathematical discoveries" over his years at *Scientific American*. Plenty of those were probably not worth taking seriously.

Gardner didn't understand what Rice sent him. He couldn't make heads or tails of her invented mathematical notation. That alone might have led another mathematician to toss Rice's work in the trash. But Gardner didn't do that. How Gardner knew that Rice's work was a diamond in the rough, we will never know. But he kept an open mind and sent it to a real tiling expert who he thought might understand it, mathematician Doris Schattschneider.

Schattschneider saw the import of what Rice had done. A less generous person might have seen the glimmers of mathematical progress in Rice's work and passed it off as her own. But Schattschneider decided to work with Rice to develop Rice's own work. Together, they presented Rice's pentagons to the world. They showed everyone who cared that mathematicians weren't finished with tiling pentagons. They still had much more work to do.

Rice and Schattschneider developed one of the richest, but also least likely, professional relationships the field of mathematics had ever seen. Altogether, Rice would find four more categories of pentagons that tile the plane. She passed away in 2017. But she lived to see mathematician Michaël Rao prove once and for all, building on Rice's discoveries, that there are exactly fifteen categories of convex pentagons that tile the plane.

Rice faced many barriers to becoming a mathematician. Lack of education, sexist expectations for what counts as women's work, and bias in the mathematical community against work that lacks the trappings of formal mathematics are just a few. And yet, with persistence and help from a few generous, open-minded mathematical insiders, Rice made her way in. Who knows what the world would believe about pentagonal tiling if she hadn't?

An earthshaking discovery, no. A solution to a problem with wide-ranging applications, also no. But it's a beautiful piece of math: her tilings show intuition; her methods reflect the interconnection of algebra and geometry; her conceptions are abstract; and undoubtedly her work is surprising. Even more importantly, this beautiful piece of math had the power to change Rice's life. If changing someone's life in this way is not a superpower, I don't know what is.

Was she a mathematical Picasso? Perhaps yes, perhaps no. But she was, as Schattschneider wrote in her obituary, "a most unlikely mathematician." The power of mathematics comes in part from the problems it solves. But the mathematicians, likely and unlikely, are ones who wield it. Math would have no power without them. Let's distribute that power widely and see what people do with it.

REFERENCES

Chapter 1. Is Math the Universal Language?

Adams, Mark. “Questioning the Inca Paradox: Did the Civilization behind Machu Picchu Really Fail to Develop a Written Language?” *Slate*, July 12, 2011. http://www.slate.com/articles/life/the_good_word/2011/07/questioning_the_inca_paradox.html.

Alex, Bridget. “Unraveling a Secret.” *Discover Magazine*, October 2017. http://discovermagazine.com/2017/oct/unraveling-a-secret.

“Arecibo Message.” SETI Institute, accessed September 1, 2019. https://www.seti.org/seti-institute/project/details/arecibo-message.

Ascher, Marcia, and Robert Ascher. *Code of the Quipu: A Study in Media, Mathematics, and Culture*. Ann Arbor: University of Michigan Press, 1981.

Busch, Michael W., and Rachel M. Reddick. “Testing SETI Message Designs.” *ArXiv*, November 20, 2009.

Collins, Allan, and William Ferguson. “Epistemic Forms and Epistemic Games: Structures and Strategies to Guide Inquiry.” *Educational Psychologist* 28, no. 1 (1993): 25–42. https://doi.org/10.1207/s15326985ep2801_3.

“Communicating across the Cosmos (Carl DeVito).” YouTube video, 00:24:57. Posted by SETI Institute, November 11, 2014. https://www.youtube.com/watch?v=MAIUVqTlDSQ&feature=youtu.be&t=3m.

Cook, Gareth. “Untangling the Mystery of the Inca.” *Wired*, January 1, 2017. https://www.wired.com/2007/01/khipu/.

Dumas, Stéphane. “The 1999 and 2003 Messages Explained.” Universe of Discourse website, accessed September 1, 2019. https://www.plover.com/misc/Dumas-Dutil/messages.pdf.

Dutil, Yvan, and Dumas Stéphane. “Annotated Cosmic Call Primer.” Smithsonian.com, September 26, 2016. http://www.smithsonianmag.com/science-nature/annotated-cosmic-call-primer-180960566/.

Franklin, K. J. “Obituary: Donald C. Laycock (1936–1988).” *Language and Linguistics in Melanesia* 20, no. 1–2 (1989): 1–5.

Freudenthal, Hans. *Lincos: Design of a Language for Cosmic Intercourse*. Amsterdam: North-Holland, 1960.

Halberstadt, Jason. “Inca Expansion and the Conquistadors.” Ecuador Explorer, accessed September 1, 2019. http://www.ecuadorexplorer.com/html/inca_expanison__the_conquista.html.

Hyland, Sabine. "Writing with Twisted Cords: The Inscriptive Capacity of Andean Khipus." *Current Anthropology* 58, no. 3 (April 19, 2017): 412–19. https://doi.org/10.1086/691682.

Keim, Brandon. "Building a Better Alien Calling Code." *Wired*, November 23, 2009. https://www.wired.com/2009/11/better-seti-code/.

Kennedy, Maev. "Mathematical Secrets of Ancient Tablet Unlocked after Nearly a Century of Study." *The Guardian*, August 24, 2017. http://www.theguardian.com/science/2017/aug/24/mathematical-secrets-of-ancient-tablet-unlocked-after-nearly-a-century-of-study.

Lamb, Evelyn. "Don't Fall for Babylonian Trigonometry Hype." *Roots of Unity* (blog), August 29, 2017. http://blogs.scientificamerican.com/roots-of-unity/dont-fall-for-babylonian-trigonometry-hype/.

Laycock, D. C. "Observations on Number Systems and Semantics." In *New Guinea Area Languages and Language Study*, Pacific Linguistics 1, ed. S. A. Wurm, 219–33. Canberra: Australian National University, 1975.

Locke, L. Leland. "The Ancient Quipu, a Peruvian Knot Record." *American Anthropologist* 14, no. 2 (1912): 325–32.

Mansfield, Daniel F., and N. J. Wildberger. "Plimpton 322 Is Babylonian Exact Sexagesimal Trigonometry." *Historia Mathematica* 44, no. 4 (November 1, 2017): 395–419. https://doi.org/10.1016/j.hm.2017.08.001.

Maor, Eli. *Trigonometric Delights*. Princeton, NJ: Princeton University Press, 1998.

"Mathematical Objects Relating to Charter Members of the MAA." Smithsonian Institution, accessed September 1, 2019. https://www.si.edu/spotlight/maa-charter/computing-devices-l-leland-locke.

Murphy, Melissa Scott. "Grave Analysis." NOVA, accessed January 9, 2018. http://www.pbs.org/wgbh/nova/inca/grav-nf.html.

———. "Puruchuco-Huaquerones Project, Peru." Bryn Mawr College Anthropology Department, accessed October 28, 2017. http://people.brynmawr.edu/msmurphy/MSMresearch.htm.

"Museo de Sitio 'Arturo Jiménez Borja'—Puruchuco." Peru Ministry of Culture, accessed January 9, 2018. http://www.visitalima.pe/index.php/museos/museo-de-sitio-arturo-jimenez-borja-puruchuco.

"Obituaries: George Arthur Plimpton." *New York History* 18, no. 3 (1937): 318–25.

O'Connor, J. J., and E. F. Robertson. "Babylonian Numerals." MacTutor History of Mathematics Archive, December 2000. http://www-history.mcs.st-and.ac.uk/HistTopics/Babylonian_numerals.html.

Roach, John. "Dozens of Inca Mummies Discovered Buried in Peru." *National Geographic News*, March 11, 2004. https://news.nationalgeographic.com/news/2004/03/0311_040311_incamummies.html.

Robson, Eleanor. "Neither Sherlock Holmes nor Babylon: A Reassessment of Plimpton 322." *Historia Mathematica* 28, no. 3 (August 1, 2001): 167–206. https://doi.org/10.1006/hmat.2001.2317.

Saxe, Geoffrey B. *Cultural Development of Mathematical Ideas: Papua New Guinea Studies*. New York: Cambridge University Press, 2012.

"Sesame Street Fish to Infinity." YouTube video, 00:02:09. Posted by Beverley Louise, August 18, 2013. https://www.youtube.com/watch?v=hgZwSRpfouQ.

Taylor, David, and Tegan Taylor. "Babylonian Tablet Plimpton 322 Will Make Studying Maths Easier, Mathematician Says." ABC News, August 25, 2017. http://www.abc.net.au/news/2017-08-25/babylonian-tablet-unlocks-simpler-trigonometry-mathematics/8841368.

Tyson, Neil deGrasse. "The Search for Life in the Universe." NASA, June 30, 2003. http://www.nasa.gov/vision/universe/starsgalaxies/search_life_I.html.

Urton, Gary, and Carrie Brezine. "Khipu Accounting in Ancient Peru." *Science* 309, no. 5737 (August 12, 2005): 1065–67.

———. "Khipu from the Site of Puruchuco." Khipu Database Project, accessed October 28, 2017. http://khipukamayuq.fas.harvard.edu/KGPuruchuco.html.

———. "What Is a Khipu?" Khipu Database Project, accessed October 28, 2017. http://khipukamayuq.fas.harvard.edu/WhatIsAKhipu.html.

Woodruff, Charles E. "The Evolution of Modern Numerals from Ancient Tally Marks." *American Mathematical Monthly* 16, no. 8/9 (September 1909): 125–33. https://doi.org/10.2307/2970818.

"The World Factbook: Papua New Guinea." Central Intelligence Agency, last updated August 26, 2019. https://www.cia.gov/library/publications/the-world-factbook/geos/pp.html.

Zorn, Paul, and Barry Cipra. "Rewriting History." *What's Happening in the Mathematical Sciences* 5 (2002): 54–59.

Chapter 2. Can Math Predict the Next Move?

"About Us." Mega Millions website, accessed September 1, 2019. http://www.megamillions.com/history-of-the-game.

Aliprantis, Charalambos D., and Subir K. Chakrabarti. *Games and Decision Making*. Oxford: Oxford University Press, 1998.

Axelrod, Robert, and William D. Hamilton. "The Evolution of Cooperation." *Science* 211, no. 4489 (March 27, 1981): 1390–96.

Beal Conjecture website, accessed September 7, 2017. http://www.bealconjecture.com.

Berlekamp, Elwyn R., John H. Conway, and Richard K. Guy. *Winning Ways for Your Mathematical Plays*. Vol. 1. 2nd ed. Natick, MA: AK Peters/CRC Press, 2001.

Bowling, Michael, Neil Burch, Michael Johanson, and Oskari Tammelin. "Heads-Up Limit Hold'em Poker Is Solved." *Science* 347, no. 628 (January 9, 2015): 145–49. https://doi.org/10.1126/science.1259433.

Burr, Mike. "Paul McCartney Heads List of World's Richest Singers." *Prefixmag* (blog), September 13, 2012. http://www.prefixmag.com/news/paul-mccartney-heads-list-of-worlds-richest-singer/69105/.

Burton, Earl. "Learning about 'The Professor, the Banker and the Suicide King.'" Poker News, accessed September 7, 2017. https://www.pokernews.com/news/2005/07/professor-banker-suicide-king.htm.

Chen, Albert C., Y. Helio Yang, and F. Fred Chen. "A Statistical Analysis of Popular Lottery 'Winning' Strategies." *CS-BIGS* 4, no. 1 (2010): 66–72.

"Coming Out Simulator." Website of Nicky Case, accessed June 13, 2018. http://ncase.me/cos/.

Conover, Emily. "Texas Hold 'Em Poker Solved by Computer." *Science*, January 8, 2015. http://www.sciencemag.org/news/2015/01/texas-hold-em-poker-solved-computer.

D'Amato, Al. "Make Online Poker Legal? It Already Is." *Washington Post*, April 22, 2011. https://www.washingtonpost.com/opinions/former-senator-alfonse-damato-make-online-poker-legal-it-already-is/2011/04/20/AFAWPwOE_story.html.

Dash, Mike. "The Story of the WWI Christmas Truce." Smithsonian.com, accessed January 7, 2018. https://www.smithsonianmag.com/history/the-story-of-the-wwi-christmas-truce-11972213/.

"18 U.S. Code §1084. Transmission of Wagering Information; Penalties." Legal Information Institute, accessed September 7, 2017. https://www.law.cornell.edu/uscode/text/18/1084.

"The Evolution of Trust." Website of Nicky Case, accessed January 8, 2018. http://ncase.me/trust/.

Fefferman Lab website, accessed January 13, 2018. http://feffermanlab.org.

Goode, Erica. "John F. Nash Jr., Math Genius Defined by a 'Beautiful Mind,' Dies at 86." *New York Times*, May 24, 2015. https://www.nytimes.com/2015/05/25/science/john- nash-a-beautiful-mind-subject-and-nobel-winner-dies-at-86.html.

Guskin, Emily, Mark Jurkowitz, and Amy Mitchell. "Network: By the Numbers." The State of the News Media 2013: An Annual Report on American Journalism. Washington, DC: Pew Research Center's Project for Excellence in Journalism, March 17, 2013. http://www.stateofthemedia

.org/2013/network-news-a-year-of-change-and-challenge-at-nbc/network-by-the-numbers/.

Henchman, Joseph. "Pretend You Won the Powerball. What Taxes Do You Owe?" *Tax Foundation* (blog), January 12, 2016. https://taxfoundation.org/pretend-you-won-powerball-what-taxes-do-you-owe/.

Howard, Ms. Gail. *Lottery Winning Strategies and 70 Percent Win Formula.* Las Vegas: Smart Luck, 2014.

Johnson, Halie. "Torrey Pines Celebrates Homecoming after Two Big Wins." *Del Mar Times*, October 10, 2010. http://www.delmartimes.net/sddmt-torrey-pines-celebrates-homecoming-after-two-big-2010oct10-story.html.

Kim, Susanna. "Mega Millions Picks $356M Numbers." ABC News, March 28, 2012. http://abcnews.go.com/Business/mega-millions-356m-lucky-numbers-19-34-44/story?id=16014648.

Kohler, Chris. "7 Most Catastrophic World of Warcraft Moments." *Wired*, December 7, 2010. https://www.wired.com/2010/12/world-of-warcraft-catastrophes/.

"List of Poker Variants." Wikipedia, accessed September 3, 2019. https://en.wikipedia.org/w/index.php?title=List_of_poker_variants&oldid=792655127.

Lofgren, Eric T., and Nina H. Fefferman. "The Untapped Potential of Virtual Game Worlds to Shed Light on Real World Epidemics." *Lancet Infectious Diseases* 7, no. 9 (September 2007): 625–29. http://dx.doi.org/10.1016/S1473-3099(07)70212-8.

Madrigal, Alexis C. "How Checkers Was Solved." *The Atlantic*, July 19, 2017. https://www.theatlantic.com/technology/archive/2017/07/marion-tinsley-checkers/534111/.

Meyer, Gerhard, Marc von Meduna, Tim Brosowski, and Tobias Hayer. "Is Poker a Game of Skill or Chance? A Quasi-Experimental Study." *Journal of Gambling Studies* 29, no. 3 (September 2013): 535–50. https://doi.org/10.1007/s10899-012-9327-8.

Minkel, J. R. "Computers Solve Checkers—It's a Draw." *Scientific American*, July 19, 2007. https://www.scientificamerican.com/article/computers-solve-checkers-its-a-draw/.

Munroe, Randall. "Tic-Tac-Toe." *XKCD* (blog), accessed August 5, 2017. https://xkcd.com/832/.

"Parable of the Polygons." Website of Nicky Case, accessed June 13, 2018. http://ncase.me/polygons.

Paradis, Bryce. "Humans, Robots, and the Consequences." *Cepheus Poker Project* (blog), January 8, 2015. http://poker-blog.srv.ualberta.ca/2015/01/08/humans-robots-and-the-consequences.html.

Sandholm, Tuomas. "Solving Imperfect-Information Games." *Science* 347, no. 6218 (January 9, 2015): 122–23. https://doi.org/10.1126/science.aaa4614.

Schaeffer, Jonathan, Neil Burch, Yngvi Björnsson, Akihiro Kishimoto, Martin Müller, Robert Lake, Paul Lu, and Steve Sutphen. "Checkers Is Solved." *Science* 317, no. 5844 (September 14, 2007): 1518–22. https://doi.org/10.1126/science.1144079.

Serna, Joseph. "So What's a Better Bet: Powerball, Mega Millions or Super Lotto Plus? We Ask the Experts." *Los Angeles Times*, July 8, 2016. http://www.latimes.com/local/lanow/la-me-ln-lottery-powerball-megamillions-odds-jackpots-20160707-snap-story.html.

Shinzaki, Michael. "I Beat the 'Unbeatable' Poker-Playing Artificial Intelligence—Sort Of." *Slate*, April 13, 2015. http://www.slate.com/blogs/future_tense/2015/04/13/i_beat_cepheus_the_unbeatable_poker_playing_artificial_intelligence.html.

Thomas, Robert M., Jr. "Marion Tinsley, 68, Unmatched as Checkers Champion, Is Dead." *New York Times*, April 8, 1995. http://www.nytimes.com/1995/04/08/obituaries/marion-tinsley-68-unmatched-as-checkers-champion-is-dead.html.

Chapter 3. Can Math Eliminate Bias?

"About." Redistricting Majority Project, accessed January 13, 2018. http://www.redistrictingmajorityproject.com/?page_id=2.

"Adult Compas Assessment: Risk and Pre-Screen." Northpointe Institute for Public Management, Inc., 2007. https://assets.documentcloud.org/documents/2840632/Sample-Risk-Assessment- COMPAS-Risk-and-Pre.pdf.

"Age of Algorithms: Data, Democracy and the News." Vimeo livestream, 01:40:01. Posted by NYU Arthur L. Carter Journalism Institute, February 15, 2017. https://livestream.com/accounts/17645697/events/7009934?t=1528484098.

Andrews, Donald A., Lina Guzzo, Peter Raynor, Robert C. Rowe, L. Jill Rettinger, Albert Brews, and J. Stephen Wormith. "Are the Major Risk/Need Factors Predictive of Both Female and Male Reoffending? A Test with the Eight Domains of the Level of Service/Case Management Inventory." *International Journal of Offender Therapy and Comparative Criminology* 56, no. 1 (2012): 113–33. https://doi.org/10.1177/0306624X10395716.

Angwin, Julia. "About." Website of Julia Angwin, accessed July 20, 2017. http://juliaangwin.com/about/.

Angwin, Julia, Jeff Larson, Lauren Kirchner, and Surya Mattu. "Minority Neighborhoods Pay Higher Car Insurance Premiums Than White Areas with the Same Risk." ProPublica, April 5, 2017. https://www.propublica.org/article/minority-neighborhoods-higher-car-insurance-premiums-white-areas-same-risk.

Angwin, Julia, Jeff Larson, Surya Mattu, and Lauren Kirchner. "Machine Bias: There's Software Used across the Country to Predict Future Criminals. And It's Biased against Blacks." ProPublica, May 23, 2016. https://www.propublica.org/article/machine-bias-risk-assessments-in-criminal-sentencing.

Angwin, Julia, Surya Mattu, and Jeff Larson. "The Tiger Mom Tax: Asians Are Nearly Twice as Likely to Get a Higher Price from Princeton Review." ProPublica, September 1, 2015. https://www.propublica.org/article/asians-nearly-twice-as-likely-to-get-higher-price-from-princeton-review.

Associated Press. "Attorney General Issues Tougher Rules under Bail Reform Law." *US News & World Report*, May 24, 2017. https://www.usnews.com/news/best-states/new-jersey/articles/2017-05-24/attorney-general-issues-tougher-rules-under-bail-reform-law.

Astor, Maggie, and K. K. Rebecca Lai. "What's Stronger Than a Blue Wave? Gerrymandered Districts." *New York Times*, November 29, 2018. https://www.nytimes.com/interactive/2018/11/29/us/politics/north-carolina-gerrymandering.html.

Barnes, Robert. "Supreme Court Takes on Texas Gerrymandering Case, Will Look at Internet Sales Tax." *Washington Post*, January 12, 2018. https://www.washingtonpost.com/politics/supreme-court-will-look-at-texas-gerrymandering-case-internet-sales-tax/2018/01/12/3744896a-f7aa-11e7-beb6-c8d48830c54d_story.html.

Blum-Smith, Ben. "Partisan Gerrymandering and Measures of Fairness." Presented at the Geometry of Redistricting, University of San Francisco, March 17, 2018.

Cebul, R. D., and R. M. Poses. "The Comparative Cost-Effectiveness of Statistical Decision Rules and Experienced Physicians in Pharyngitis Management." *Journal of the American Medical Association* 256 (1986): 3353–57.

Cho, Wendy K. Tam, and Yan Y. Liu. "Sampling from Complicated and Unknown Distributions: Monte Carlo and Markov Chain Monte Carlo Methods for Redistricting." *Physica A: Statistical Mechanics and Its Applications* 506 (September 15, 2018): 170–78. https://doi.org/10.1016/j.physa.2018.03.096.

———. "Toward a Talismanic Redistricting Tool: A Computational Method for Identifying Extreme Redistricting Plans." *Election Law Journal* 15, no. 4 (2016): 351–66. https://doi.org/10.1089/elj.2016.0384.

Christian, Brian, and Tom Griffiths. *Algorithms to Live By: The Computer Science of Human Decisions*. New York: Picador, 2016.

"Congress and the Public." Gallup.com, accessed August 1, 2017. http://www.gallup.com/poll/1600/Congress-Public.aspx.

D'Agostino, Susan. "Sharing Math's Appeal with First-Generation Students." *Chronicle of Higher Education*, October 28, 2013. http://chronicle.com/article/Designing-a-Math-Major-That/142551/.

Dieterich, William, Christina Mendoza, and Tim Brennan. *COMPAS Risk Scales: Demonstrating Accuracy Equity and Predictive Parity. Performance of the COMPAS Risk Scales in Broward County*. Traverse City, MI: Northpointe, Inc., July 8, 2016.

"Drawing the Lines on Gerrymandering." CBS News, January 14, 2018. https://www.cbsnews.com/news/drawing-the-lines-on-gerrymandering/.

Duchin, Moon, and Peter Levine. "Rebooting the Mathematics behind Gerrymandering." The Conversation, October 23, 2017. http://theconversation.com/rebooting-the-mathematics-behind-gerrymandering-73096.

Ferguson, Thomas S. "Who Solved the Secretary Problem?" *Statistical Science* 4, no. 3 (August 1989): 282–89.

Foderaro, Lisa W. "New Jersey Alters Its Bail System and Upends Legal Landscape." *New York Times*, February 6, 2017. https://www.nytimes.com/2017/02/06/nyregion/new-jersey-bail-system.html?_r=0.

Fushimi, Masanori. "The Secretary Problem in a Competitive Situation." *Journal of the Operations Research Society of Japan* 24, no. 4 (December 1981): 350–58.

"Gerrymandering." YouTube video, 00:19:33. Posted by Last Week Tonight with John Oliver, April 9, 2017. https://www.youtube.com/watch?v=A-4dIImaodQ.

"Gerrymandering and Partisan Politics in the U.S." PBS NewsHour, September 26, 2016. http://www.pbs.org/newshour/extra/daily_videos/gerrymandering-and-partisan-politics-in-the-u-s/.

Grove, William M., David H. Zald, Boyd S. Lebow, Beth E. Snitz, and Chad Nelson. "Clinical Versus Mechanical Prediction: A Meta-Analysis." *Psychological Assessment* 12, no. 1 (2000): 19–30. https://doi.org/10.1037//1040-3590.12.1.19.

Haselhurst, Geoff. "Bertrand Russell Quotes on Mathematics/Mathematical Physics." On Truth and Reality, accessed July 20, 2017. http://www.spaceandmotion.com/mathematical-physics/famous-mathematics-quotes.htm.

"History of Federal Voting Rights Laws: The Voting Rights Act of 1965." US Department of Justice, accessed June 8, 2018. https://www.justice.gov/crt/history-federal-voting-rights-laws.

"How Elections Are Rigged—Gerrymandering." YouTube video, 01:17:17. Posted by SnagFilms, September 21, 2016. https://www.youtube.com/watch?v=-285T7Pdp58.

Klarreich, Erica. "Gerrymandering Is Illegal, but Only Mathematicians Can Prove It." *Wired*, April 16, 2017. https://www.wired.com/2017/04/gerrymandering-illegal-mathematicians-can-prove/.

Kleinberg, Jon, Sendhil Mullainathan, and Manish Raghavan. "Inherent Trade-Offs in the Fair Determination of Risk Scores." *Proceedings of Innovations in Theoretical Computer Science*, 2017. http://arxiv.org/abs/1609.05807.

Larson, Jeff, Surya Mattu, Lauren Kirchner, and Julia Angwin. "How We Analyzed the COMPAS Recidivism Algorithm." ProPublica, May 23, 2016. https://www.propublica.org/article/how-we-analyzed-the-compas-recidivism-algorithm.

Liptak, Adam. "Justices Reject 2 Gerrymandered North Carolina Districts, Citing Racial Bias." *New York Times*, May 22, 2017. https://www.nytimes.com/2017/05/22/us/politics/supreme-court-north-carolina-congressional-districts.html.

———. "Supreme Court Bars Challenges to Partisan Gerrymandering." *New York Times*, June 27, 2019. https://www.nytimes.com/2019/06/27/us/politics/supreme-court-gerrymandering.html.

"Minnesota 2016 Population Estimates." US Census Bureau, accessed July 28, 2017. https://www.census.gov/quickfacts/MN.

"New Jersey Eliminates Cash Bail, Leads Nation in Reforms." PBS NewsHour, July 22, 2017. http://www.pbs.org/video/3003039348/.

"North Carolina Congressional Redistricting after the 2010 Census." Carolina Demography, accessed July 21, 2017. http://www.arcgis.com/apps/StorytellingSwipe/index.html?appid=a15c27c984ed404782da753dd840e99a.

"North Carolina's 11th Congressional District." Ballotpedia, accessed July 21, 2017. https://ballotpedia.org/North_Carolina's_11th_Congressional_District.

"North Carolina 2018 Population Estimates." US Census Bureau, accessed July 29, 2017. https://www.census.gov/quickfacts/NC.

Parker, Matt. "The Secretary Problem." *Slate*, December 17, 2014. http://www.slate.com/articles/technology/technology/2014/12/the_secretary_problem_use_this_algorithm_to_determine_exactly_how_many_people.html.

Pearson, Rick. "Federal Court Approves Illinois Congressional Map." *Chicago Tribune*, December 16, 2011. http://articles.chicagotribune.com/2011-12-16/news/ct-met-congress-map-court-20111216_1_congressional-map-earmuff-shaped-new-map.

———. "Federal Court Upholds Democrats' Map of Illinois Congressional Districts." *Los Angeles Times*, December 15, 2011. http://www.latimes.com

/nation/politics/politicsnow/chi-federal-court-upholds-democrats-map -of-illinois-congressional-districts-20111215-story.html.

"Public Safety Assessment." Laura and John Arnold Foundation, accessed July 26, 2017. http://www.arnoldfoundation.org/initiative/criminal-justice /crime-prevention/public-safety-assessment/.

"Public Safety Assessment: Risk Factors and Formula." Laura and John Arnold Foundation, accessed July 26, 2017. http://www.arnoldfoundation .org/wp-content/uploads/PSA-Risk-Factors-and-Formula.pdf.

"REDMAP: How a Strategy of Targeting State Legislative Races in 2010 Led to a Republican U.S. House Majority in 2013." Redistricting Majority Project, January 4, 2013. http://www.redistrictingmajorityproject.com /?p=646.

Rosenblum, Dan. "Hakeem Jeffries Gives the Prison-Gerrymander Presentation at His Old Law School." *Politico*, January 24, 2012. http://www.politico .com/states/new-york/city-hall/story/2012/01/hakeem-jeffries-gives-the -prison-gerrymander-presentation-at-his- old-law-school-069480.

Rowlett, Russ. "Names for Large Numbers." How Many? A Dictionary of Units of Measurement, accessed July 28, 2017. https://www.unc.edu /~rowlett/units/large.html.

Tippett, Rebecca. "Redistricting North Carolina in 2011." *Carolina Demography* (blog), November 10, 2015. http://demography.cpc.unc.edu/2015/11/10 /redistricting-north-carolina-in-2011/.

"20 David Kung, Empowering Who? The Challenge of Diversifying the Mathematical Community." YouTube video, 01:03:46. Posted by Educational Advancement Foundation, July 21, 2015. https://www.youtube.com /watch?v=V03scHu_OJE.

"2010 Census Tallies." US Census Bureau, accessed July 28, 2017. https:// www.census.gov/geographies/reference-files/time-series/geo/tallies.html.

"2011 Redistricting Process." North Carolina General Assembly, accessed July 21, 2017. http://www.ncleg.net/representation/Content/Process2011 .aspx.

"2012 House Election Results by Race Rating." Cook Political Report, November 9, 2012. http://cookpolitical.com/house/charts/race-ratings /5123.

"2012 REDMAP Summary Report." Redistricting Majority Project, January 4, 2013. http://www.redistrictingmajorityproject.com/?p=646.

Vieth v. Jubelirer. 541 U.S. 267 (2004).

"Wendy Cho: Enabling Redistricting Reform. A Computational Study of Zoning Optimization." YouTube video, 00:24:03. Posted by NCSAatIllinois, June 12, 2017. https://www.youtube.com/watch?v=-OSGq6zOejw.

Wines, Michael, and Richard Fausset. “North Carolina Is Ordered to Redraw Its Gerrymandered Congressional Map. Again.” *New York Times*, August 30, 2018. https://www.nytimes.com/2018/08/27/us/north-carolina-congressional-districts.html.

Chapter 4. Can Math Open Doors?

Alleyne, Ayinde, Ben Blum-Smith, and Lynn Cartwright-Punnet. BEAM staff interview with author. July 31, 2015.

Blum-Smith, Ben. “Kids Summarizing.” *Research in Practice* (blog), September 8, 2013. https://researchinpractice.wordpress.com/2013/09/08/kids-summarizing/.

“Brahman, Caste.” *Encyclopedia Britannica Online*, accessed November 10, 2017. https://www.britannica.com/topic/Brahman-caste.

Burr, Stefan A., and George E. Andrews. *The Unreasonable Effectiveness of Number Theory*. Providence, RI: American Mathematical Society, 1993.

Caldwell, Chris K. “Database Search Output (Another of the Prime Pages' Resources).” Prime Pages, accessed September 3, 2019. http://primes.utm.edu/primes/search.php?Comment=twin%20OR%20triplet&Number=100.

Cook, Gareth. “The Singular Mind of Terry Tao.” *New York Times*, July 24, 2015. https://www.nytimes.com/2015/07/26/magazine/the-singular-mind-of-terry-tao.html.

Cook, Mariana. *Mathematicians: An Outer View of the Inner World*. Princeton, NJ: Princeton University Press, 2009.

Cooke, Roger L. *The History of Mathematics: A Brief Course*. 3rd ed. Hoboken, NJ: John Wiley & Sons, 2013.

“Demographic Snapshot 2012–13 to 2016–17 Public—Citywide, Borough, District, and School.” NYC Department of Education, April 12, 2017. http://schools.nyc.gov/Accountability/data/default.htm.

Douglass, Frederick. *Frederick Douglass, Autobiographies: Narrative of the Life of Frederick Douglass, an American Slave/My Bondage and My Freedom/Life and Times of Frederick Douglass*. Edited by Henry Louis Gates. New York: Library of America, 1994.

Dumas, Michael J. “‘Losing an Arm’: Schooling as a Site of Black Suffering.” *Race Ethnicity and Education* 17, no. 1 (2014): 1–29. https://doi.org/10.1080/13613324.2013.850412.

“Enrollment—Statewide by Institution Type.” Texas Higher Education Coordinating Board, 2016. http://www.txhighereddata.org/index.cfm?objectid=867CFDB0-D279-6B64-55037383E42EE290.

"Ilana." Video of student interview with BEAM group. Summer 2015.

Jimenez, Laura, Scott Sargrad, Jessica Morales, and Maggie Thompson. *Remedial Education: The Cost of Catching Up*. Washington, DC: Center for American Progress, September 2016. https://cdn.americanprogress.org/content/uploads/2016/09/29120402/CostOfCatchingUp2-report.pdf.

Kanigel, Robert. *The Man Who Knew Infinity*. New York: Washington Square Press, 1991.

Moses, Robert P., and Charles E. Cobb Jr. *Radical Equations: Civil Rights from Mississippi to the Algebra Project*. Boston: Beacon Press, 2001.

Nasir, Na'ilah Suad, Cyndy R. Snyder, Niral Shah, and Kihana Miraya Ross. "Racial Storylines and Implications for Learning." *Human Development* 55, no. 5–6 (2012): 285–301. https://doi.org/10.1159/000345318.

National Science Board. *Science and Engineering Indicators 2012*. NSB 12-01. Arlington, VA: National Science Foundation, 2012.

Chapter 5. What Is Genuine Beauty?

Bolyai, János. "The Science Absolute of Space." *Scientiae Baccalaureus* 1, no. 4 (June 1891).

Cannon, James W., William J. Floyd, Richard Kenyon, and Walter Parry. "Hyperbolic Geometry." *Flavors of Geometry* 31 (1997): 59–115.

Casey, John, and Euclid. *The First Six Books of the Elements of Euclid*. Dublin: Hodges, Figgis, 1885. https://www.gutenberg.org/files/21076/21076-pdf.pdf.

Cook, Mariana. *Mathematicians: An Outer View of the Inner World*. Princeton, NJ: Princeton University Press, 2009.

Creativity Research Group website, accessed January 14, 2018. http://www.creativityresearchgroup.com/.

Delp, Kelly, Craig S. Kaplan, Douglas McKenna, and Reza Sarhangi, eds. *Bridges Baltimore: Mathematics, Music, Art, Architecture. Culture Conference Proceedings*. Phoenix, AZ: Tessallations, 2015. http://archive.bridgesmathart.org/2015/frontmatter.pdf.

Demaine, Erik, Martin Demaine, and Anna Lubiw. "Hyperbolic Paraboloids." Erik Demaine's Folding and Unfolding, last updated May 28, 2014. http://erikdemaine.org/hypar/.

Dénes, Tamás. "The Real Face of Janos Bolyai." *Notices of the American Mathematical Society* 58, no. 1 (January 2011): 41–51.

"Euclid's Elements: Book 1." Website of David E. Joyce, accessed September 3, 2019. https://mathcs.clarku.edu/~djoyce/elements/bookI/bookI.html.

"Euclid's Elements: Book 1, Definition 1 Guide." Website of David E. Joyce, accessed September 3, 2019. https://mathcs.clarku.edu/~djoyce/elements/bookI/defI1.html.

"Eve Torrence." Mathematical Art Galleries website, accessed September 3, 2019. http://gallery.bridgesmathart.org/exhibitions/2015-bridges-conference/etorrenc.

Frazer, Jennifer. "Proteus: How Radiolarians Saved Ernst Haeckel." *Artful Amoeba* (blog), January 31, 2012. https://blogs.scientificamerican.com/artful-amoeba/proteus-how-radiolarians-saved-ernst-haeckel/.

Gardner, Martin. "Mathematical Games: On Tessellating the Plane with Convex Polygon Tiles." *Scientific American* 233, no. 1 (1975): 112–19.

Hardy, G. H. *A Mathematician's Apology*. Electronic ed., version 1.0. Edmonton: University of Alberta Mathematical Sciences Society, 2005. https://www.math.ualberta.ca/mss/misc/A%20Mathematician%27s%20Apology.pdf.

Hartshorne, Robin. *Geometry: Euclid and Beyond*. Undergraduate Texts in Mathematics. New York: Springer, 2000.

Henderson, David W., and Daina Taimina. "Crocheting the Hyperbolic Plane." *Mathematical Intelligencer* 23, no. 2 (March 1, 2001): 17–28. https://doi.org/10.1007/BF03026623.

——. "Experiencing Meanings in Geometry." In *Mathematics and the Aesthetic: New Approaches to an Ancient Affinity*. Canadian Mathematical Society Books in Mathematics. New York: Springer Science + Business Media B. V., 2006.

Kandinsky, Wassily. *On the Spiritual in Art: First Complete English Translation, with Four Full Colour Page Reproductions, Woodcuts and Half Tones*. New York: Solomon R. Guggenheim Foundation, 1946. https://archive.org/details/onspiritualinart00kand.

——. *Point and Line to Plane: Contribution to the Analysis of the Pictorial Elements*. New York: Solomon R. Guggenheim Foundation, 1947. https://archive.org/details/pointlinetoplane00kand.

Kershner, R. B. "On Paving the Plane." *American Mathematical Monthly* 75, no. 8 (October 1, 1968): 839–44. https://doi.org/10.1080/00029890.1968.11971075.

——. "On Paving the Plane." *APL Technical Digest* (July 1969): 4–10.

Lamb, Evelyn, and Kevin Knudson. "Mohamed Omar's Favorite Theorem." *Roots of Unity* (blog), January 11, 2018. https://blogs.scientificamerican.com/roots-of-unity/mohamed-omars-favorite-theorem/.

Lewis, Florence P. "History of the Parallel Postulate." *American Mathematical Monthly* 27, no. 1 (January 1920): 16–23.

Loomis, Elisha S. *The Pythagorean Proposition*. Washington, DC: National Council of Teachers of Mathematics, 1968.

Marshall, Daniel, and Paul Scott. "A Brief History of Non-Euclidean Geometry." *Australian Mathematics Teacher* 60, no. 3 (March 2004): 2–4.

Movshovits-Hadar, Nitsa. "School Mathematics Theorems: An Endless Source of Surprise." *For the Learning of Mathematics* 8, no. 3 (November 1988): 34–40.
"The Nature of Things/Martin Gardner." Vimeo video, 00:46:05. Posted by Wagner Brenner, October 20, 2009. https://vimeo.com/7176521.
Nelsen, Roger B. *Proofs without Words: Exercises in Visual Thinking*. Washington, DC: Mathematical Association of America, 1993.
"Quotations." Mathematical Association of America, accessed January 5, 2018. https://www.maa.org/press/periodicals/convergence/quotations.
Rice, Marjorie. "Tessellations—Intriguing Tessellations." Tessellations website, accessed February 21, 2018. https://sites.google.com/site/intriguingtessellations/home/tessellations.
Schattschneider, Doris. "Marjorie Rice (16 February 1923–2 July 2017)." *Journal of Mathematics and the Arts* 12, no. 1 (November 28, 2017): 51–54. https://doi.org/10.1080/17513472.2017.1399680.
Sinclair, Nathalie, and David Pimm. "A Historical Gaze at the Mathematical Aesthetic." In *Mathematics and the Aesthetic: New Approaches to an Ancient Affinity*. Canadian Mathematical Society Books in Mathematics. New York: Springer Science + Business Media B. V., 2006.
Strogatz, Steven. *The Joy of X: A Guided Tour of Math, from One to Infinity*. Boston: Houghton Mifflin Harcourt, 2012.
Tao, Terence. "Pythagoras' Theorem." *What's New* (blog), September 14, 2007. https://terrytao.wordpress.com/2007/09/14/pythagoras-theorem/.
Torrence, Eve. Email interview with author. March 31, 2017.
Weisstein, Eric W. "Playfair's Axiom." Wolfram MathWorld, accessed January 6, 2018. http://mathworld.wolfram.com/PlayfairsAxiom.html.
Wolchover, Natalie. "Pentagon Tiling Proof Solves Century-Old Math Problem." *Quanta Magazine*, July 11, 2017. https://www.quantamagazine.org/pentagon-tiling-proof-solves-century-old-math-problem-20170711/.

INDEX